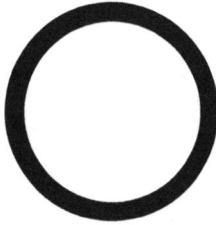

First established in 2004, the DATA browser book series explores new thinking and practice at the intersection of contemporary art, digital culture and politics. The series takes theory or criticism not as a fixed set of tools or practices, but rather as an evolving chain of ideas that recognize the conditions of their own making. The term "browser" is useful here in pointing to the framing device through which data is delivered over information networks and processed by algorithms. Whereas a conventional understanding of browsing suggests surface readings and cursory engagement with the material, the series celebrates the potential of browsing for dynamic rearrangement and interpretation of existing material into new configurations that are open to reinvention.

Series editors:

Geoff Cox
Joasia Krysa

Volumes in the series:

DB 01 ECONOMISING CULTURE
DB 02 ENGINEERING CULTURE
DB 03 CURATING IMMATERIALITY
DB 04 CREATING INSECURITY
DB 05 DISRUPTING BUSINESS
DB 06 EXECUTING PRACTICES
DB 07 FABRICATING PUBLICS
DB 08 VOLUMETRIC REGIMES

www.data-browser.net

This volume is produced with support from Liverpool John Moores University and London South Bank University.

DATA browser 08
VOLUMETRIC REGIMES:
Material cultures of
quantified presence

Possible Bodies
Blanca Pujals
Jara Rocha
Femke Snelting
Sina Seifee
Nicolas Malevé
Simone C Niquille
Maria Dada
Phil Langley
Helen V. Pritchard
Kym Ward
Spec
Sophie Boiron
Pierre Huyghebaert
Romi Ron Morrison
The Underground Division
Manetta Berends

DATA browser 08
VOLUMETRIC REGIMES: Material
cultures of quantified presence

Edited by Possible Bodies
volumetricregimes.xyz

Published by
Open Humanities Press 2022
Copyright ⓩ 2022 the authors

PDF freely available at
data-browser.net/db08.html

ISBN (print): 978-1-78542-116-7
ISBN (PDF): 978-1-78542-115-0

DATA browser series template
designed by Stuart Bertolotti-Bailey

Wiki-to-print development and
F/LOSS redesign by Manetta Berends
Source files: git.vvvvvvaria.org/mb/
volumetric-regimes-book

The cover image is derived from
Variable Geometry, an Open Source
and cross-platform interpretation by
Winnie Soon and Geoff Cox of the
software app Multi by David Reinfurt.
Multi updates the idea of the multiple
from industrial production to the
dynamics of the information age.
Each cover presents an iteration of a
possible 1,728 arrangements, each a
face built from minimal typographic
furniture, and from the same source
code. o-r-g.com/apps/multi
and aesthetic-programming.net/
pages/2-variable-geometry.html

Contents

Acknowledgements

Volumetric Regimes is the outcome of a collective process, inhabiting a thick network of para-academic solidarity between practitioners of different media, methods and tongues. We first of all would like to thank the interlocutors that have contributed to this book with their wonderful thinking, drawing, imagining and writing: Sophie Boiron, Maria Dada, Pierre Huyghebaert, Phil Langley, Nicolas Malevé, Romi Ron Morrison, Simone C Niquille, Helen V. Pritchard, Blanca Pujals, Sina Seifee and Kym Ward.

A huge thanks also to those who have been in conversation with the project at large, including: Ramon Amaro, Mercé Ardèvol, Tere Badia, Laura Benítez, Ona Bros, Gonzalo Correa, Emile Devereaux, Daphne Dragona, Marta Echaves, Laura Fernández, Sonia Fernández-Pan, Abelardo Gil-Fournier, Antye Guenther, Seda Gürses, Marie Lechner, Max and Franz Lehner, Alejandra López Gabrielidis, Zoumana Meïté, Martino Morandi, Paula Pin, Dennis Pohl, Anna Ramos, Carmen Romero Bachiller, SpiderAlex, Sophie Toupin, Peter Westenberg, Kathryn Yusoff, François Zajega and Adva Zakai.

The articulation of this research became a possibility during a fellowship at Akademie Schloss Solitude (Science and Business department) in Stuttgart and was further developed with a grant from the Flemish Government between 2017 and 2018. Other cultural and academic organisations supported us by inviting Possible Bodies for workshops, exhibitions, discussions and residencies: transmediale, Berlin; Hangar, Barcelona; Medialab Prado, Madrid; Constant, Brussels; Furtherfield, London; Jan van Eyck Academie, Maastricht; Festival Gelatina, Madrid; Universidad de la República and Casa Mario, Montevideo; Fuga, Barcelona; Goldsmiths University, London; Museo Reina Sofía, Madrid; La Gaîté Lyrique, Paris; Het Nieuwe Instituut, Rotterdam; Radio MACBA, Barcelona; a.pass, Brussels; Fem TEK, Bilbao; FAP-TEK, Montevideo; University of Sussex, Brighton; Azala, Lasierra; and CSNI, London. Thank you participants, colleagues, comrades and also organizers, editors and curators with whom we had long and short exchanges over the years.

We want to acknowledge everyone who helped make this work available, accountable and legible, from on-line hosting, designing, peer-reviewing and transcribing to copy-editing. Thank you Geoff Cox and Joasia Krysa for your generous support as editors; Constant for providing us with an array of tools and practices plus a wide window to display our tilted sensibilities; Manetta Berends for the inspiring design collaboration and your comradeship; Nerea Calvillo, Eric Snodgrass, Magda Tyżlik-Carver for your invaluable comments combining rigour with enthusiasm; Mara Ittel, Fanny Wendt Höjer and Marc Herbst for your meticulous attention to detail and a special thanks to Helen V. Pritchard for their close companionship in the ongoing revolving of all matters.

We are deeply grateful for all the maintenance and care work involved in the making of first the artistic research of the Possible Bodies Inventory, and into a book later. *Volumetric Regimes* would not exist without the encouraging and supportive energies coming from companions, colleagues, friends, ancestors, and lovers of many sorts. Thank you all, for the inspiring and groundbreaking questions that you kept asking, full of constructive critique and sharp provocations. The depths and densities we need to embrace complexity with, would certainly not be possible without these thick currents of radical interdependencies.

Foreword

Blanca Pujals

We need, I believe, to engage a different kind of violence, a violence that is neither spectacular nor instantaneous, but rather incremental and accretive, its calamitous repercussions playing out across a range of temporal scales.[1]

```
#!/usr/env/python3
import numpy as np
import matplotlib.pyplot as plt
from mpl_toolkits.mplot3d import Axes3D
fig = plt.figure()
ax = fig.add_subplot(111, projection='3d')
ax.set_aspect('equal')
u = np.linspace(0, 2 * np.pi, 100)
v = np.linspace(0, np.pi, 100)
x = 1 * np.outer(np.cos(u), np.sin(v))
y = 1 * np.outer(np.sin(u), np.sin(v))
z = 1 * np.outer(np.ores(np.size(u)), np.cos(v))
elev = 10.0
rot = 80.0 / 180 * np.pi
ax.plot_surface(x, y, z, rstride=4, cstride=4, color='b', linewidth=0, alpha=0.5)
a = np.array([-np.sin(elev / 180 * np.pi), 0, np.cos(elev / 180 * np.pi)])
b = np.array([0, 1, 0])
b = b * np.cos(rot) + np.cross(a, b) * np.sin(rot) + a * np.dot(a, b) * (1 - np.cos(rot))
ax.plot(np.sin(u),np.cos(u),0,color='k', linestyle = 'dashed')
horiz_front = np.linspace(0, np.pi, 100)
ax.plot(np.sin(horiz_front),np.cos(horiz_front),0,color='k')
vert_front = np.linspace(np.pi / 2, 3 * np.pi / 2, 100)
ax.plot(a[0] * np.sin(u) + b[0] * np.cos(u), b[1] * np.cos(u), a[2] * np.sin(u) + b[2] * np.cos(u),color='k', linestyle = 'dashed')
```

A Sphere. Coded by psy (03c8.net)

The design of so-called bodies, territories or organisms, and of narratives which outline *difference* and *The Other*, impose binary separations. They produce slow a-temporal resonances throughout time, systematically replicating a western epistemology of management and control. These constructions allowed the contemporary techno-scientific management of the environment and of our bodies, producing and reproducing algorithmic and systemic discrimination, which reveal different forms of structural differences and hegemonic fictions.

With-technology, "we" emerge as "us". A hybrid human-machine-electron-organism, a planetary/time-space where cables, rare earth, bodies, soil, entities, liquids, particles, atmosphere and outer space are increasingly interconnected in a techno-organism that is speedily evolving, but which gathers in its source code a historical continuum of feedback loops.[2]

In his book *Slow Violence*, Rob Nixon describes processes characterized by violence that occur gradually and often in invisible ways.[3] Perhaps we can say that the parametric disassembling and reassembling of bodies and territories, is a process of slow violence that brings these problems to the present through their uninterrupted and increasingly sophisticated implementation.

The scientific revolution, understood as a sociotechnical momentum in which the values of Modernity where implemented across western science disciplines, brought a transformation of the way the world was understood, starting off a massive taxonomization and abstraction of the environment. With processes of fragmentation, repeatability and simulation, the splitting of behaviors and organisms into data permeated into the volumetric design of bodies and territories. Since then, we find them embedded in computational software architectures, technologies which relentlessly scan matter in search of new forms of intervention. Hence, so-called bodies, territories, organisms, the organic and the inorganic can be managed for an efficient extraction and manipulation of materials and data, unveiling different forms of possession, property, rights and conflicts.

Statistical techniques of averaging and correlation from the nineteenth century introduced new photographic methods into their analyses, to catalogue organs and matter into static and isolated units. The pictures were organized and categorized by similar units in comparative tables, as for example Alphonse Bertillon's *bertillonage*, or Francis Galton's *composites*, where images, as slices, were overlapped to reveal a standard for disease, criminal type or race. Reading *Volumetric Regimes*, we can see how these techniques, initiated by the science of criminology, are nowadays introduced in the form of scanning and modelling technologies to disassemble and reassemble reality. Thus, these *us*-devices construct artificial borders, isolating groups of archetypes in a desperate will to contain nature's behavior, analyzed and fixed through parametric systems into physical and digital technoscientific containment architectures. This provides a fantasy of enclosure, preventing the release and spread of an-Other's influence. A system based on behavioral speculation and probability that creates both new threats and objectives within a retroactive system: a knowledge of the future by probabilistic determinism that continuously feeds a feedback loop of fears and threats, which shapes and reshapes our entire sociality.

In *Volumetric Regimes* we find, as a kind of resonance chamber full of case studies, an inventory of techniques used in the context of 3D computing to artificially design *humanness*, referred to as so-called bodies, so-called earth or so-called plants. Mechanisms such as rigging, agential cuts, slicing, dividing, dimensional axes of power, x,y,z, simulated environments, processes of modelling, capturing, rendering, printing and tracking unveil how scientific knowledge incorporated in computational tools is still based on dividing, separating and creating boundaries in a fictional composition of the tangible, in which the world is bounded and organized according to categories of hegemonic fictions. Through this organization, objects and organisms are disintegrated within extreme division and classification, and the isolation of the parts for their analysis erodes the uncategorized interrelations. Imposed official landscapes and bodies disregard the intra-actions "as a dynamism of forces in which all designated 'things' are constantly exchanging and diffracting, influencing and working inseparably."[4]

Nevertheless, as the book also points out, new results concerning the origin of matter can lead us to reconsider older categories. The Heisenberg uncertainty principle states that one cannot simultaneously know both the exact position and exact momentum of a single particle. In addition, quantum entanglement means that the quantum state of each particle of a group cannot be described independently of the state of the others, including when particles are separated by a large distance, and so, also measurements affect the entangled system as a whole.[5] In this picture, matter moves increasingly toward a hybrid, ungraspable state. Quantum physics defies the system of observation born with the Modern project. Therefore, concepts such as uncertainty or entanglement, fundamental properties of the very elementary particles of matter (paradoxically resulting from splitting and smashing them to build the provisional Standard Model), can provide us with new generative possibilities and forms of social imagination.

Although liminal matter problematizes the matter of facts, "bodies" are still, and increasingly, locked within a regime of Modernity. As this book shows, this regime quietly and systematically spreads through contemporary 3D computing. Possible Bodies explores in *Volumetric Regimes* what the imaginary produced within that ontological and epistemological status of computational volumetrics does, and how it intervenes into power relationships. At the same time, they offer us a new imagine-action to rethink previous categorizations, by renaming them.[6] "Languaging" is one of the main tactics for unsettling Modern assumptions and rigidities. Vocabularies, verbalizations and discourse articulations provide with a rich realm for suspending the probable and extending the spectrum of what's possible. "In other words", language is understood as a mode for keeping complexity close while not aligning with the manners and grammars of a damaging world setting.

Perhaps, following the bug reports included in *Volumetric Regimes*, we can try to rethink the semantic layer of computational processes. As the image at the beginning of this text suggests, at the basis of computational design, we have mathematical code, a text which, as Possible Bodies proposes, we might need to re-imagine and re-write from scratch.

Notes

1. ↑ Rob Nixon, *Slow Violence and the Environmentalism of the Poor* (Harvard: Harvard University Press, 2011).

2. ↑ "Feedback occurs when outputs of a system are routed back as inputs as part of a chain of cause-and-effect that forms a circuit or loop. The system can then be said to feed back into itself. Simple causal reasoning about a feedback system is difficult because the first system influences the second and second system influences the first, leading to a circular argument." "Feedback," *Wikipedia*, accessed October 28, 2021, https://en.wikipedia.org/wiki/Feedback.

3. ↑ Nixon, *Slow Violence and the Environmentalism of the Poor.*

4. ↑ Karen Barad, *Meeting the Universe Half Way* (Durham: Duke University Press, 2007).

5. ↑ Quantum Entanglement, Wikipedia, accessed November 1, 2021, https://en.wikipedia.org/wiki/Quantum_entanglement.

6. ↑ Imagin-action (Imagina-ação): imagination as intervention in reproductive systems and invention of worlds; The "faculty", or the operation of imagination in its collective and performative character of intervention in the world and not as a mere abstraction or pure speculative activity. In Portuguese, the word for "imagination" is "imaginação" and it contains the word "action", "ação". In this sense, thinking about the activity of imagining as an action, "ImaginAction" is an ethical, ontoepistemic and pragmatic approach, since it can lead us to other ways of conceiving the activity of imagining. Amilcar Packer, Office of Political Imagination, Brazil, 2016-now.

Volumetric Regimes: Material cultures of quantified presence

Possible Bodies (Jara Rocha, Femke Snelting)

What is going on with 3D!? This question, both modest and enormous, triggered the collaborative research trajectory that is compiled in this book. It was provoked by an intuitive concern about the way 3D computing quite routinely seems to render racist, sexist, ableist, speciest and ageist worlds.[1] Asking about what is up with 3D becomes especially urgent given its application in border-patrol devices, for climate prediction modeling, in advanced biomedical imaging or throughout the gamify-all approach of overarching industries, from education to logistics. The proliferating technologies, infrastructures and techniques of 3D tracking, modeling and scanning are increasingly hard to escape.

Volumetric Regimes emerges from Possible Bodies, a collaborative, multi-local and polyphonic project situated on the intersection of artistic and academic research, developing alongside an inventory of cases through writing, workshops, visual essays and performances. This publication brings together diverse materials from that trajectory as well as introduces new materials. It represents a rich and ongoing conversation between artists, software developers and theorists on the political, aesthetic and relational regimes in which volumes are calculated. At some point, we decided to fork Possible Bodies into The Underground Division to name the intensifying conversations on 3D-geocomputation with Helen V. Pritchard.[2] This explains why the attribution of the materials compiled in *Volumetric Regimes* takes multiple expressions of an extended *we*.[3]

When we asked "What is going on with 3D?!" we generated many further questions, such as: Why is *3D* now used as a synonym for *volume-metrics*. Or: how did the metric of volume become naturalized as *3D*? How are volumes computed, accounted for and represented? Is the three-dimensional technoscientific organization of spaces, bodies or objects only about volume, or rather about the particular modes in which volume is culturally mobilized? How, then, are computational volumes occupying the world? What forms of power come along with 3D? How are the x, y and z axes established as linear carriers or variables of volume, by whom and why? If we take 3D as a noun, it points at the quality of being three-dimensional. But what if we follow the intuition of asking about "what is going on" and take 3D as an action, as an operation with implications for the way we can world otherwise? Can 3D be turned into a verb, at all? How can we at the same time use, problematize and engage with the cultures of volume-processing that converge under the paradigm of 3D?

One important question we almost overlooked: What is volume, actually!? Let's start by saying that as a representation of mass and of matter, volume is a naturalized construction, by means of calculation. "3D" then is a shortcut for the cultural means by which contemporary volume gets produced, especially in the context of computation. By persistently foregrounding its three distinct dimensions: depth, height and width, the concept of volume gets inextricably tied to the Cartesian coordinate system, a particular way of measuring dimensional worlds. The cases and situations compiled in this book depart from this important shift: in computation, volume is not a given, but rather *an outcome*, and volumetrics is the set of techniques to fabricate such outcome.

As a field oriented towards the technocratic realm of Modern technosciences, 3D computation has historically unfolded under "the probable" regimes of optimization, totalitarian efficiency, normalization and hegemonic world order.[4] Think of scanning the underground for extractable petro-fossil resources with the help of technologies first developed for brain surgery, and large scale agro-industrial 3D-applications such as spray installations enhanced with fruit recognition. In that sense, volumetrics is involved in sustaining the all too probable behavior of 3D, which is actively being (re)produced and accentuated by digital hyper-computation. The legacies and projections of industrial development leave traces of an ongoing controversy, where multiple modes of existence become increasingly unimaginable under the regime of the probable. *Volumetric Regimes* explores operational, discursive and procedural elements which might widen "the possible" in contemporary volumetrics.

Material cultures

This book is an inquiry into the material cultures of volumetrics. We did not settle on one specific area of knowledge, but rather stayed with the complexity of intricate stories that in one way or another involve a metrics of volume. The study of material cultures has a long tail which connects several disciplines, from archaeology and ethnography or design, which each bring their own methodological nuances and specific devices. *Volumetric Regimes* sympathizes with this multi-fold research sensibility that is necessary to think-with-matter. The framework of material cultures provides us with an arsenal of tools and vocabularies, interlocuting with, for example, New Feminist Materialisms, Science and Technology Studies, Phenomenology, Social Ecology or Cultural Studies.

The study of the material cultures of volumetrics necessitates a double-bind approach. The first bind relates to the material culture of volume. We need to speak about the volume that so-called bodies occupy in space from the material perspective of what they are made of – the actual conditions of their material presence and the implications of what space they occupy, or not. But we also need to speak about the material arrangements of metrics, the whole ecology of tools that participate in measuring operations. The second bind is therefore about the technopolitical aspects of knowledge production by measuring matter and of measured matter itself; in other words: the material culture of metrics.

The material culture of volumetrics and its internal double bind implies an understanding of technosocial relations as always in the making, both shaping and being shaped under the conditions of cultural formations. Being sensitive to matter therefore also involves a critical accountability towards the exclusions, reproductions and limitations that such formations execute. We decided to approach this complexity by assuming our response-ability with an inventory filled with cases and an explicitly political attitude.

The way matter matters has a direct affect on how something becomes a structural and structured regime, or rather how it becomes an ongoing contingent amalgamation of forces. There is no doubt that metrics can be considered to be a cultural realm of its own,[5] but what about the possibility of volume as a cultural field, infused by an apparatus of axioms and assumptions that, despite their rigid affirmations, are not referring to a pre-existent reality, but actually rendering one of their own?

In this book, we spend some quality time with the idea that volume, is the product of a specific evolution of material culture. We want to activate a public conversation, asking: How is power distributed in a world that is worlded by axes, planes, dimensions and coordinates, too often and too soon crystallizing abstractions in a path towards naturalizing what presences count where, for whom and for how long?

Volumetric regimes

We started this introduction by saying that volume is an outcome, not a given. Mass can (but does not have to) be measured by culturally-set operations like the calculation of its depth, or of its density. The volumes resulting from such measurement operations use cultural or scientific assumptions such as limit, segment or surface. The specific ways that volumetrics happen, and the modes that result in them crystallizing into axes and axioms, are the ones that we are trying to trace back and forth, to identify how they ended up arranging a whole regime of thought and praxis.

The contemporary regime of volumetrics, meaning the enviro-socio-technical politics and narratives that emerge with and around the measurement and generation of 3D presences, is a regime full of bugs, crawling with enviro-socio-technical flaws. Not neutral and also not innocent, this regime is wrapped up in the interrelated legacies and ideologies of neoliberalism, patriarchal colonial commercial capitalism, tied with the oligopolies of authoritarian innovation and technoscientific mono-cultures of proprietary hardware and software industries, intertwined with the cultural regimes of mathematics, image processing but also overly rigid vocabularies. In feminist techno-science, the relation between (human) bodies and technologies has had lots of attention, from the cyborg manifesto to more recent new materialist renderings of phenomena and apparatuses.[6] In the field of software studies, the "deviceful" entanglements between hegemonic regimes and software procedures have been thoroughly discussed,[7] while anti-colonial scholars have critiqued the ways that measuring or metrics align with racial capitalism and North-South divisions of power.[8] Thinking about the computation of volume is merely present in literature on the interaction of human and other-than-human bodies with machinic agents,[9] with the built environment[10] and its operative logics.[11]

What we have been looking for in the works listed above, and not always found, is the kind of diffuse rigor needed for a transformative politics that is a condition for non-binarism, of not settling, of being response-able in constant change.[12] This search triggered the intense interlocutions with the artists, activists and thinkers that have contributed to this book, and made us stick to polyhedric research methods. We've gone back to Paul B. Preciado, who taught us about the political fiction that so-called bodies are, a fleshy accumulation of archival data that keeps producing, reproducing and/or contesting the truths of power and their interlinked subjectivities.[13] Fired up for the worlding of *different tech,* we found inspiring unfoldings of computation and geological volumes in Kathryn Yusoff's and Elizabeth A. Povinelli's work, who insist on brave unpackings of Modern regimes all-the-way. We wondered about the voluminosity of "bodies" but also about their entanglement with what marks them as such, and how to pay attention to it. Reading Denise Fereirra da Silva's email conversation with Arjuna Neuman about her use of the term "Deep Implicancy" rather than "entanglement", we were struck by the relation between spatiality and separation she brings up: "Deep Implicancy is an attempt to move away from how separation informs the notion of entanglement. Quantum physicists have chosen the term entanglement precisely because their starting point is particles (that is, bodies), which are by definition separate in space."[14] Syed Mustafa Ali and David Golumbia separate computation from computationalism to make clear that while computation obviously sediments and continues colonial damages, this is not necessarily how it needs to be (and it necessarily needs to be otherwise). Interlocutions with the deeply situated work of Seda Gürses,[15] operating on the discipline of computation from the inside, sparked with the energy of queer thinkers and artists Zach Blas and Micha Cárdenas[16] and more recently Loren Britton, and Helen V. Pritchard in *For CS.*[17] We are grateful for their critical problematizations of the ever-straightening protocols which operate in every corner where existence is supposed to happen.

The shift to understanding volume as an outcome of sociotechnical operations, is what helps us activate the critical revision of the regimes of volumetry and their many consequences. If volume does not exist without volumetric regimes, then the technopolitical struggle means to scrutinize how metrics could be exploded, (re)designed, otherwise implemented, differently practiced, (de)bugged, interpreted and/or cared for.

Quantified presence

Volumetric Regimes is also our way to build capacities for a response to the massive quantification of presences existing in computed space-times. Such response-ability needs to be multi-faceted, due to the process of manipulation that quantifying presences apply to presence itself as an ontological concern. The fact that something can exist and be accountable in a virtual place, or that something which is present in a physical space can re-appear or be re-presented in differently mediated conditions, or not at all, is technically produced through supposedly efficient gestures such as clear-cut incisions, separating boundaries, layers of segmentation, regions of interest and acts of discretization. The agency of these operations is more often than not erased after the fact, providing a nauseating sense of neutrality.

The project of *Volumetric Regimes* is to think with and towards computing-otherwise rather than to side with the uncomputable or to count on that which escapes calculation. Flesh, complexity and mess are already-with computation, somehow simultaneous and co-constituent of mess. The spaces created by the tension between matter and its quantification, provide with a creative arena for the diversification of options in the praxis of 3D computation. Qualitative procedures like intense dialog, hands-on experiments, participant observation, speculative design and indeterminate protocols help us understand possible research attitudes in response to a quantify-all monoculture, not succumbing to its own pre-established analytics. Could "deep implicancy" be where computing otherwise happens, by means of speculation, indeterminacy and possibility? Perhaps such praxis is already located beyond or below normed actions like capturing, modeling or tracking that are all so complicit with the making of fungibility.[18]

The specific form of quantification that is at stake in the realm of volume-metrics, is datafication. The computational processing, displacing and re-arranging of matter through volumetric techniques participates in what The Invisible Committee called *the crisis of presence*, which can be observed at the very core of the contemporary ethos.[19] We connect with their concerns about the way present presences are rendered, or not. How to value what needs to count and be counted or what is in excess of quantification, via the exact same operation, in a politicized way. In other words, a politics of reclaiming quantification is a praxis towards a politicized accountability for the messiness of all techniques that deal with the thickness of a complex world. Such praxis is not against making cuts as such, but rather commits to being response-able with the gestures of discretion and not making final finite gestures, but reviewable ones. Connecting to quantification in this manner, is a claim for forms of accountable accountability.[20]

Aligning ourselves with the tradition of feminist techno-sciences, *Volumetric Regimes: Material cultures of quantified presence* stays with the possible (possible tools, methods, practices, material-izations, agencies, vocabularies) of computation, demanding complexity while queering the rigidity of their fixing of items, discrete and finite entities in too fast moves towards truth and neutrality.

In this publication we try by all means to disorient the assumption of essentialist discreteness and claims for the thickening of qualitative presence in 3D computation realms. In that sense, *Volumetric Regimes* could be considered as an attempt to do qualitative research on the quantitive methods related to the volumetric-occupation of worlds.

Polyhedric research methods

In terms of method, this book benefits from several polyhedric forces that when combined, form a prismatic body of disciplinarily uncalibrated but rigorous research. The study of the complex regimes that rule the worlds of volumes, necessitated a few methodological inventions to widen the spectrum of how computational volumetrics can be studied, described, problematized and reclaimed.[21] That complexity is generated not only by the different areas in which measuring volumes is done, but also because it is a highly crowded field, populated by institutional, commercial, scientific, sensorial and technological agents.

One polyhedric force is the need for direct action and informed disobedience applied to research processes. We have often referred to our work as "disobedient action-research", to insist on a mode of research that is motivated by situated, ad-hoc modes of producing and circulating knowledge. We committed to a non-linear workflow of writing, conversing and referencing, to keep resisting developmental escalation, but rather to hold on to an iterative and sometimes recursive flow. While in every discipline there are people and practices opening, mixing, expanding, challenging, and refusing traditional methods, research involving technology is too often ethically, ontologically, and epistemologically dependent on a path from and towards universalist enlightenment, aiming to eventually technically fixing the world. This violent and homogenizing solutionist attitude stands in the way of a practice that, first of all, needs to attend to the re-articulation and relocation of what must be accounted for, perhaps just by proliferating sensibilities, issues, demands, requests, complaints, entanglements, and/or questions.[22]

A second polyhedric force is generated by the playful intersection of artistic and academic research in the collaborative praxis of Possible Bodies. It materializes for example in uncommon writing and the use of made-up terminology, but also in the hands-on engagement with tools, merging high and low tech, learning on the go, while attending to genealogies that arranged them in the here-now. You will find us smuggling techniques for knowledge generation from one domain to another such as contaminating ethnographic descriptions with software stories, mixing poetics with abnormal visual renders, blurring theoretical dissertations with industrial case-studies and so forth.

Trans*feminism is certainly a polyhedric dynamic at work, in mutual affection with the previous forces. We refer to the research as such, in order to convoke around that star (*) all intersectional and *intra-sectional* aspects that are possibly needed.[23] Our trans*feminist lens is sharpened by queer and anti-colonial sensibilities, and oriented towards (but not limited to) trans*generational, trans*media, trans*disciplinary, trans*geopolitical, trans*expertise, and trans*genealogical forms of study. The situated mixing of software studies, media archaeology, artistic research, science and technology studies, critical theory and queer-anticolonial-feminist-antifa-technosciences purposefully counters hierarchies, subalternities, privileges and erasures in disciplinary methods.

The last polyhedric force is generated by our politicized attitude towards technological objects. This book was developed on a wiki, designed with Free, Libre and Open Source software (FLOSS) tools and published as Open Access.[24] Without wanting to suggest that FLOSS itself produces the conditions for non-hegemonic imaginations, we are convinced that its persistent commitment to transformation can facilitate radical experiments, and trans*feminist technical prototyping. The software projects we picked for study and experimentation such as Gplates,[25] MakeHuman,[26] and Slicer[27] follow that same logic. It also oriented our DIWO attitude of investigation, preferring low-tech approaches to high-tech phenomena and allowing ourselves to misuse and fail.

To give an ongoing account of the structural formations conditioning the various cultural artifacts that are co-composed through scanning, tracking and modeling, we settled for **inventorying** as a central method. The items in the Possible Bodies inventory do not rarefy these artifacts, as would happen through the practice of collecting, or by pinning them down, as in the practice of cartography, or rigidly stabilize them, as might be a risk through the practice of archiving.[28] Instead, the inventorying is about continuous updates, and keeping items available. The inventory functions as an additional reference system for building stories and vocabularies; items have been used for multiple guided tours, both written and performed.[29] Being aware of its problematic histories of commercial colonialism, the praxis of inventorying needs to also be reoriented towards just and solidary techniques of semiotic-material compilation.[30]

The writing of **bug reports** is a specific form of disobedient action research which implies a systematic re-learning of the very exercise of writing, as well as a resulting direct interpellation to the communities that develop software, by its own means and channels. Bug reporting, as a form of technical grey literature, makes errors, malfunctions, lacks, or knots legible; secondly, it reproduces a culture of a public interest in actively taking-part in contemporary technosciences. As a research method, it can be understood as a repoliticization and cross-pollination of one of the key traditional pillars of scientific knowledge production: the publishing of findings.

Technical expertise is not the only knowledge suitable for addressing the technologically produced situations we find ourselves in. The term **clumsy computing** describes a mode of relating to technological objects that is diffuse, sensitive, tentative but unapologetically confident.[31] Such diffuseness can be found in the selection of items in the inventory,[32] in the deliberate use of deported terminology, in the amateur approach to tools, in the hesitation towards supposedly ontologically-static objects of study, in the sudden scale jumps, in the radical disciplinary un-calibration and in our attention to porous boundaries of sticky entities.[33]

The persistent use of **languaging formulas** problematizes the limitations of ontological figures. For example the repeated use of "so-called" for "bodies" or "plants" is a way to question the various methods whereby finite, specified and discrete entities are being made to represent the characteristics of whole species, erasing the nuances of very particular beings.[34] Combinatory terms such as "somatopologies" play a recombinatory game to insist on the implications of one regime onto another.[35] Turning nouns into verbs such as using "circlusion" as "circluding", is a technology that forces language to operate with different temporary tenses and conjugations, refusing the fixed ontological commingling that naming implies.[36]

Interlocution has ruled the orientations of this inquiry that was collective by default: by affecting and being affected by communities of concern in different locations, the research process changed perspectives, was infused by diverse vocabularies and sensibilities and jumped scales all along. The conversations brought together in *Volumetric Regimes* stuck with this principle of developing the research through an affective network of comrades, companions, colleagues and collaborators, based on elasticity and mutual co-constitution.

README

Volumetric Regimes experiments with various formats of writing, publishing and conversing. It compiles guided tours, peer-reviewed academic texts, speculative fiction, pamphlets, bug reports, visual essays, performance scripts and inventory items. It is organized around five chapters, that each rotate the proliferating technologies, infrastructures and techniques of 3D tracking, modeling and scanning differently. Although they each take on the question "What is going on with 3D?!" through a distinct *axiology* of technology, politics and aesthetics, they do not assume nor impose a specific order for the reader. Each chapter includes an invited contribution that proposes a different orientation, offers a Point of View (POV) or triggers a perspective on the material-discursive entanglements in its own way.

x, y, z: Dimensional Axes of Power takes on the building blocks of 3D: x, y and z. The three Cartesian axes both constrain and orient the chapter, as they do for the space of possibility of the volumetric. It takes seriously the implications of a mathematical regime based on parallel and perpendicular lines, and zooms in on the invasive operations of virtual renderings of fleshy matter, but also calls for queer rotations and disobedient trans∗feminist angles that can go beyond the rigidness of axiomatic axes within the techno-ecologies of 3D tracking, modeling and scanning. The chapter begins with a contribution by Sina Seifee, who in his text "Rigging Demons" draws from an intimate history with the technical craft-intense practice of special effects animation, to tell us stories of visceral non-mammalian animality between love and vanquish. The chapter continues with a first visit to the Possible Bodies inventory that sets-up the basic suspicions on what is of value in rendered and captured worlds, following the thread of dis-orientation as a way to think through the powerful worldings that are nevertheless produced by volumetrics. "Invasive

Imagination and its agential cut" reflects on the regimes of biomedical imaging and the volumetrization of so-called bodies.

Somatopologies: On the ongoing rendering of corpo-realities opens up all the twists in epistemologies and methodologies triggered by *Volumetric Regimes* in the somatic realm. As a notion, "somatopologies" converges the not-letting-go of Modern patriarcho-colonial apparatuses of knowledge production like mathematics or geometry, specifically focusing on an undisciplined study of the paradigm of topology. By opening up the conditions of possibility, somatopologies is a direct claim for other ontologies, ethics, practices and crossings. The chapter opens with "Clumsy Volumetrics" in which Helen V. Pritchard follows Sara Ahmed's suggestion that "clumsiness" might form a queer and crip ethics that generates new openings and possibilities. "Somatopologies (materials for a movie in the making)" documents a series of installations and performances that mixed different text sources to cut agential slices through technocratic paradigms in order to create hyperbolic incisions that stretch, rotate and bend Euclidean nightmares and Cartesian anxieties. "Circluding" is a visual/textual collaboration with Kym Ward on the potential of a gesture that flips the order of agency without separating inside from outside. In "From Topology to Typography: A romance of 2.5D", Sophie Boiron and Pierre Huyghebaert open up a graphic conversation on the almost-volumetrics that precede 3D in digital typography and finally the short text "MakeHuman" and the pamphlet "Information for Users" take on the implications of relating to 3D-modelled-humanoids.

The vibrating connections between hyper-realism and invention, re-creation and simulation, generation and parametrization are the inner threads of a chapter titled *Parametric Unknowns: Hypercomputation between the probable and the possible*. What's in the world and what is processed by mechanisms of volumetric vision differs only slightly, offering a problematic dizzying effect. The opening of the chapter is in the hands of Nicolas Malevé, who offers a visual ethnography of some of the interiors and bodies that made computational photography into what it became. Not knowing everything yet, the panoramization of intimate atmospheres works as an exercise to study the limits between the flat surfaces of engineering labs and the dense worlds behind their scenes. "The Fragility of Life" is an excuse to enter into the thick files compiled by designer-researcher Simone C Niquille on the digital post-production of truth. Somehow in line with that, Maria Dada provides an overview of how different training and rehearsing are, especially in the gaming industry that makes History with a capital H. And finally, a long-term conversation with Phil Langley questions the making of too fast computational moves while participating in architectural and infrastructural materializations.

Signs of Clandestine Disorder: The continuous aftermath of 3D-computationalism follows the long tail of volumetric techniques, technologies and infrastructures, and the politics inscribed within. The chapter's title points to "computationalism", a direct reference to Syed Mustafa Ali's approach to decolonial computing.[37] The other half is a quote from Alphonso Lingis, which invokes the non-explicit relationality between elements that constitute computational processes.[38] In that sense, it contrasts directly with the discursive practice of colonial perception that Ramon Amaro described as "self maintaining in its capacity to empirically self-justify."[39] The chapter opens with "Endured Instances of Relation" in which Romi Ron Morrison reflects on specific types of fixity and fixation that pertain to volumetric regimes, and the radical potential of "flesh" in data practices, while understanding bodies as co-constructed by their inscriptions, as a becoming-with technology. The script for the workshop "Signs of clandestine disorder for the uniformed and codified crowd" is a generative proposal to apply the mathematical episteme to lively matters, but without letting go of its potential. In "So-called Plants" we return to the inventory for a vegetal trip, observing and describing some operations that affect the vegetal kingdom and volumetrics.

The last chapter is titled *Depths and Densities: Accidented and dissonant spacetimes*. It proposes to shift from the scale of the flesh to the scale of the earth. The learnings from the insurgent geology of authors like Yusoff triggered many questions about the ways technopolitics cut the vertical and horizontal axis and that limit the spectrum of possibilities to a universalist continuation of extractive modes of existence and knowledge production. The contribution by Kym Ward, "Open Boundary Conditions", offers a first approach to her situated intensive study of the crossings between volumetrics and oceanography, from the point of view of the Bidston Observatory in Liverpool. From this vantage point she articulates a critique on technosciences, and provides with an overview of possible affirmative areas of study and engagement. In "A Bugged Report", the filing of bug reports turns out to be an opportune way to react to the embeddedness of anthropocentrism in geomodeling software tools, different to, for example, technological sovereignty claims. "We Have Always Been Geohackers" continues that thinking and explores the probable continuation of extractive modes of existence and knowledge production in software tools for rendering tectonic plates. The workshop script for exercising an analog LiDAR apparatus is a proposal to experience these tensions in physical space, and then to discuss them collectively. The chapter ends with "Ultrasonic Dreams of Aclinical Renderings", a fiction that speculates with hardware on the possibilities for scanning through accidented and dissonant spacetimes.

Notes

1. ↑ This intuition surfaced in *GenderBlending*, a worksession organized by Constant, association for art and media based in Brussels, in 2014. Body hackers, 3D theorists, game activists, queer designers and software feminists experimented at the contact zones of gender and technology. Starting from the theoretical and material specifics of gender representations in a digital context, GenderBlending was an opportunity to develop prototypes for modelling digital bodies differently. "Genderblending," Constant, accessed October 6, 2021, https://constantvzw.org/site/-GenderBlending,190-.html.

2. ↑ "The Underground Division," accessed October 20, 2021, http://ddivision.xyz.

3. ↑ We decided not to flatten or erase these porous attributions. Therefore you will find a multiplicity of authorial takes throughout this book: Possible Bodies, Possible Bodies (Jara Rocha, Femke Snelting), Possible Bodies feat. Helen V. Pritchard, Jara Rocha, and Femke Snelting, Kym Ward feat. Possible Bodies, Jara Rocha, The Underground Division and The Underground Division (Helen V. Pritchard, Jara Rocha, Femke Snelting).

4. ↑ See "Item 127: El Proyecto Moderno / The Modern Project," *The Possible Bodies Inventory*, 2021.

5. ↑ See, for example Alfred W. Crosby, *The Measure of Reality: Quantification in Western Europe, 1250–1600* (Cambridge: Cambridge University Press, 1997).

6. ↑ Karen Barad, *Meeting the Universe Halfway* (Durham: Duke University Press, 2007).

7. ↑ Some of the publications in the field of Software Studies that have done this work include Matthew Fuller, *Behind the Blip: Essays on the Culture of Software* (Brooklyn: Autonomia, 2003), Adrian Mackenzie, *Cutting Code: Software and Sociality* (New York: Peter Lang, 2006), Wendy Chun, *Programmed Visions: Software and Memory* (Cambridge MA: MIT Press, 2011), Matthew Fuller, and Andrew Goffey, *Evil Media* (Cambridge MA: MIT Press, 2012), Geoff Cox, and Alex McLean, *Speaking Code: Coding as Aesthetic and Political Expression* (Cambridge MA: MIT Press, 2012) and more recently Winnie Soon, and Geoff Cox *Aesthetic Programming* (London: OHP, 2020).

8. ↑ From Wendy Hui Kyong Chun, "Race and/as Technology; or, How to Do Things to Race," Camera Obscura 70, 24(1) 7-35, to Ruha Benjamin,

Race After Technology: Abolitionist Tools for the New Jim Code (Hoboken: Wiley, 2019).

9. ↑ Stamatia Portanova, *Moving Without a Body* (Cambridge MA: MIT Press, 2012).

10. ↑ Luciana Parisi, *Contagious Architecture: Computation, Aesthetics, and Space* (Cambridge MA: MIT Press, 2013).

11. ↑ Aud Sissel Hoel, and Frank Lindseth, "Differential Interventions: Images as Operative Tools," in *Photomediations: A Reader*, eds. Kamila Kuc and Joanna Zylinska (Open Humanities Press, 2016), 177-183.

12. ↑ "There are no solutions; there is only the ongoing practice of being open and alive to each meeting, each intra-action, so that we might use our ability to respond, our responsibility, to help awaken, to breathe life into ever new possibilities for living justly." Karen Barad, *Meeting the universe halfway*.

13. ↑ Paul B. Preciado calls the fictive accumulation a *somathèque.* "Interview with Beatriz Preciado, SOMATHEQUE. Biopolitical production, feminisms, queer and trans practices," Radio Reina Sofia, July 7, 2012, https://radio.museoreina sofia.es/en/somatheque-biopolitical -production-feminisms-queer-and-trans-practices.

14. ↑ Denise Ferreira da Silva, and Arjuna Neuman, "4 Waters: Deep Implicancy" (Images Festival, 2019) h ttp://archive.gallerytpw.ca/wp-conte nt/uploads/2019/03/Arjuna-Denise-web-ready.pdf.

15. ↑ For example in her important work on understanding shifts in the practice of software production. Seda Gürses, and Joris Van Hoboken, "Privacy after the Agile Turn," eds. Jules Polonetsky, Omer Tene, and Evan Selinger, Cambridge Handbook of Consumer Privacy (Cambridge University Press, 2018), 579-601.

16. ↑ Zach Blas, and Micha Cárdenas, "Imaginary Computational Systems: Queer technologies and transreal aesthetics," *AI & Soc* 28 (2013): 559–566.

17. ↑ Loren Britton, and Helen Pritchard, "For CS," *interactions* 27, 4 (July - August 2020), 94–98.

18. ↑ See: Romi Ron Morrison, "Endured Instances of Relation, an exchange," in this book.

19. ↑ The Invisible Committee, *To Our Friends* (Los Angeles: Semiotext(e), 2015).

20. ↑ Karen Barad, *Meeting the Universe Halfway.*

21. ↑ Celia Lury, and Nina Wakeford, *Inventive Methods: the Happening of the Social* (Milton Park: Routledge, 2013).

22. ↑ See: The Underground Division (Helen V. Pritchard, Jara Rocha, and Femke Snelting), "We have always been geohackers," in this book.

23. ↑ "The asterisk hold off the certainty of diagnosis." Jack Halberstam, *trans∗: A Quick and Quirky Account of Gender Variability* (Berkeley: University of California Press, 2018), 4.

24. ↑ See: Manetta Berends, "The So-Called Lookalike," in this book.

25. ↑ See: Jara Rocha, "Depths and Densities: A bugged report," in this book.

26. ↑ See: Jara Rocha, Femke Snelting, "MakeHuman," in this book.

27. ↑ See: Jara Rocha, Femke Snelting, "Invasive Imagination and its Agential Cuts," in this book.

28. ↑ See: Jara Rocha, Femke Snelting, "Disorientation and its aftermath," in this book.

29. ↑ Possible Bodies, "Inventorying as a method," *The Possible Bodies Inventory,* https://possiblebodies.con stantvzw.org/inventory/?about.

30. ↑ See: Jara Rocha, Femke Snelting, "Disorientation and its aftermath," in this book.

31. ↑ See: Helen Pritchard, "Clumsy Computing", in this book.

32. ↑ *The Possible Bodies Inventory*, accessed October 20, 2021, https://possiblebodies.constantvzw.org/inventory

33. ↑ Andrea Ballestero, "The Underground as Infrastructure? Water, Figure/Background Reversals and Dissolution in Sardinal," ed. Kregg Hetherington, *Infrastructure, Environment and Life in the Anthropocene* (Durham: Duke University Press, 2019).

34. ↑ See: "So-called Plants," in this book.

35. ↑ See: "Somatopologies," in this book.

36. ↑ See: Kym Ward feat. Possible Bodies, "Circluding," in this book.

37. ↑ Syed Mustafa Ali, "A Brief Introduction to Decolonial Computing," *XRDS: Crossroads, The ACM Magazine for Students*, 22(4) (2016): 16-21.

38. ↑ "We walk the streets among hundreds of people whose patterns of lips, breasts, and genital organs we divine; they seem to us equivalent and interchangeable. Then something snares our attention: a dimple speckled with freckles on the cheek of a woman; a steel choker around the throat of a man in a business suit; a gold ring in the punctured nipple on the hard chest of a deliveryman; a big raw fist in the delicate hand of a schoolgirl; a live python coiled about the neck of a lean, lanky adolescent with coal-black skin. Signs of Clandestine Disorder in the Uniformed and Coded Crowds." Alphonso Lingis, *Dangerous Emotions* (University of California Press, 2000), 141.

39. ↑ Ramon Amaro, "Artificial Intelligence: warped, colorful forms and their unclear geometries," in *Schemas of Uncertainty: Soothsayers and Soft AI*, eds. Danae Io and Callum Copley (Amsterdam: PUB/Sandberg Instituut, 2019), 69-90.

x, y, z: Dimensional Axes of Power

Rigging Demons

Sina Seifee

An efficient particle-based rig for self-attractive dispersal nCloth objects in the 3D software Maya. Sina Seifee, 2020

Coming from the pirate infrastructure of Iran, computer black-market by default, sometime in my early youth I installed a cracked version of Maya (3D software developed at that time by Alias Wavefront). I was making exploratory locomotor behaviors, scripting postural coordinations, kinesthetic structures, and automated skeletal rigs. Soon after, doing simple computer graphics hacks in 3D became a pragmatic experimentation habit. Now looking back, I think it was a way for me to extend a line of flight. Doing autonomous affective pragmatic experiments in a virtual microworld helped me to exit my form of subjectivity. Something that I will unpack in the following text as *counter dispossession through engagement with the phantom limb*.

"Counter" is perhaps not quite the right word, *play* is more accurate. Because play happens always on the edge of double bind experience (a condition of schizophrenia). Our relationship with media technologies is a "double bind patterning", a system of layered contradictions that is experienced as reality. Following Katie King's rereading of her teacher Gregory Bateson, double bind happens when something is prohibited at one level of meaning or abstraction (within a particular communicating channel), while something else is required (at another level) that is impossible to effect if the prohibition is honored.[1] Our relationship with the phantom limb is at once experienced at the level of terror (being haunted by it) and companionship (extend one's being in the world).

Disintegration of a demon in *Charmed*, season 1, episode 20, © Warner Bros. Entertainment Inc, 1998. Collage: Sina Seifee

This text develops a system of references and compositional attunement to a technical craft-intense practice called *rigging* in computer graphics. My aim is to apply the idea of volumetric regimes to rigging, and its media specificities, as one style of animating volumetric bodies particularly naturalized in the animation industry and its technoculture. I will highlight one of its occurrences in film, namely the visual effects that are associated with disintegration of "demons" in the TV-series *Charmed* and will propose the disintegrating demon body as a multi-sited loci of meaning. Multi-sites require inquiries in more than one location, also combining different types of location: geographical, digital, temporal, and also demonological. Disintegrating demons are less interesting as a subject for analogies of body politics and more as an object of computerized zoomorphic experimentations. They are performed in specific ways in digital circumstances, which I refer to as *doing demons*.

I am going to take myself as an empirical access point to think about the ecology of practices[2] or the ecology of minds[3] that involve computerized animated nonhumans, and arrest my digital memories as a molecular material history, in order to share my sensoria among species that shape our relationships with machines. This text is also an exercise in accounting for my own *technoperceptual habituations*. The technoperceptual can refer to the assemblages of thoughts, acts of perception and of consumption that I am participating with—a term I learnt from Amit Rai in his fabulous research on the technological cultures of hacking in India.[4]

Charmed soap operatic analytics

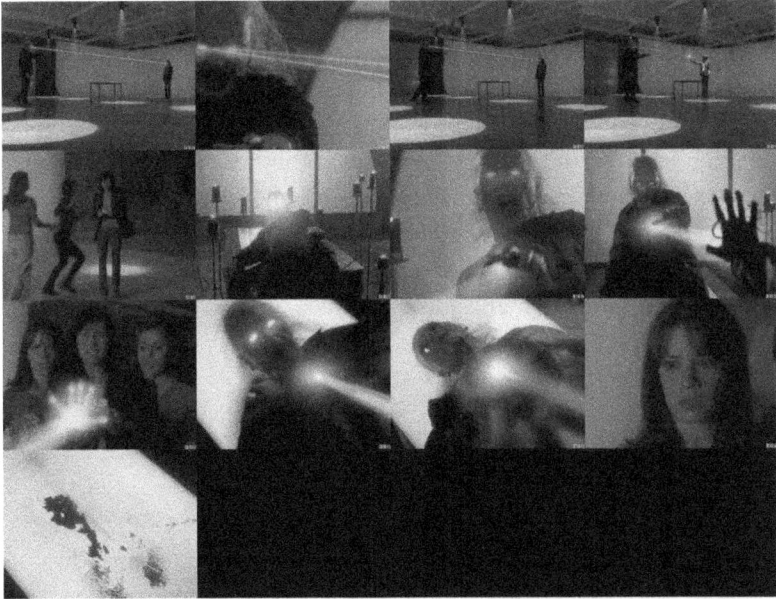

Disintegration of a demon in *Charmed* season 1, episode 2, © Warner Bros. Entertainment Inc, 1998. Collage: Sina Seifee

I was recently introduced to a multimedia franchise called *Charmed*. Broadcast by Warner Bros. Television (aired between 1998 and 2006), the adaption of *Charmed* for television is a supernatural fantasy soap opera, mixing stories of relations between women and machinic alignments. Faced with the cognitive chaos of a hypermodern life in an imaginary San Francisco, as main characters of the soap, the three sisters-witches deal with questions of narcissism (self-oriented molar life-style), prosthesis (sympathetic magic as new technologies they have to learn to live with without mastering), global networks (teamwork with underworld), and dissatisfaction (nothing works out, relationships fail, anxiety attacks, and loneliness). In the series, ancient forms for life-sources, characterized as "demons", are differentiated and encountered via the mediation of a technical life-source,

33

characterized as "magic spells". The technology is allegorically replaced by magic.

The soap presents the sisters, Prue, Phoebe, and Piper, oscillating between demon love and demon hate, and constantly negotiating the strange status of desire in general. These negotiations are fabled as the ongoing tensions between *hedonism* (to refuse to embody anxiety for polyamorous sexual life) and *tolerance* (the recognition of difference in the demons they must fight to the death) and those tensions are typically worked out melodramatically by the standards of the genre in the 1990s. The characters are frequently wrapped in and unwrapped by emotional turmoil, family discord, marriage breakdown, and secret relationships. They often show minimal interest in magic as a subject of curiosity, and instead they are more interested in spells as a medium through which their demons are externally materialized and enacted. Knowing has no effect on the protagonists' process of becoming only actions. As such *Charmed* insists on putting "the transformation of being and the transformation of knowing out of sync with one another".[5]

Past techniques of making species visible
The demons of *Charmed* are particularly interesting for multiple reasons. First, they are proposed taxonomically. Every demon is particular in its type, or subspecies, and classified per episode by its unique style of death. The demons are often mean-spirited aliens (men in suits), are less narrated in their process of becoming, and rather interested more in the classification of the manner of vanquishing them. They are "vanquished" at the end of each episode. To be more precise, exactly at minute 39, a demon is spectacularly exploded, melted, burned, or vaporized. One of the byproducts of this strange way of relating, is the *Book of Shadows*, a list or catalogue of demons and their *transmodification*. Lists are qualitative characteristics of cosmographical knowledge and my favorite specialized archival technology.

As a premodern cutting-edge agent of sorting, list-making was highly functional in the technologies of writing in the 12th and 16th century, namely monster literature, *histoire prodigieuse* or *bestiaries*. I have been thinking about bestiaries these past years, as one of the older practices of discovery, interpretation, production of the real itself. Starting off as a research project about premodern zoology in West Asia, Iran in particular, I found myself getting to know more about how "secularization of the interest in monsters"[6] happened through time. Bestiaries are *synthesized sensitized lists of the strange*. In them the enlisted creatures do not need to "stick together" in the sense of an affective or syntagmatic followability. That means they are not related narratively, but play non-abstract categories in their relentless particularities. A creative form of demon literacy, mnemonically oriented (to aid memorization), which is materialized in *Charmed* as the *Book of Shadows*. The melodrama affect of the series and emphatic lense on the love life of its cast-ensemble, allows a form of distance, making the demons becoming ontologically boring, which is paradoxically the subject of wonder literature (simultaneously distanced and intimate). On one hand the categorical nature of demons are anatomically and painfully indexed in the series, and on the other hand the romantic qualities of demonic life is explored.

Disintegration of a demon in *Charmed*, season 1, episode 22, © Warner Bros. Entertainment Inc, 1998. Collage: Sina Seifee

Soap operas are among the most effective forms of linear storytelling in the 20th century, an invention of the US daytime serials. Characteristic of a soap operatic approach, is the use of the cast-ensemble, a collective of (often glamorous and wealthy) individuals who "play off each other rather than off reality".[7] This allows the reality in which the stories go through to be rendered as an ordinary, constant, and natural stage. The soap often produces (and capitalizes on) a fable of reality, as that is the environment where multiple agencies are characteristically coordinated to face each other rather than their environment. Through the creation of banal and ordinary sites of getting on collectively in a romantic life, soup opera series are perhaps among the best tools to create cognitive companions (fans) and the sensation of ordinary affects, which are essential in "worlding" (production of the ordinary sense of a world).

Disintegration Effect on self by Surfaced Studio in "After Effects Expression Controls Tutorial - Visual Effects 101", 2012, https://youtu.be/jslSJNtoNcg

The second reason to become interested in *Charmed* demons, is because of its visual effects. The disintegration effects of *Charmed* demon vanquishing can be perceived as "low tech", meaning that its images develop a visuality that does not immediately integrate into high-end media in 2021. Its images, as I watched them in my attentive recognition (of a phenomena that is not complying with expectations) and partial attunement (to its explicit intensities), they cultivate my vision as the result of a perceiving organ. Why do I find demon species that depend on "expired" visualization technologies more interesting? This can be due to my own small resistance against new-media. Not a critical positioning, but more a sensation that has sedimented into an aesthetic taste (that is my consumption habit). The particular simulacral space of contemporary mediascape, with its preference for immersion, viscerality, interactivity, and hyperrealism, has to do with the way new-media makes meaning more *attractive* and (in a

Deleuzian sense) less *intensive*. *Charmed*'s mythopoetic dreamscape now in 2021 has lost its "appeal", therefore it is available to become tasty again. A witness to the gain and loss of attractivity in media culture is the process of fixing "bad" visual effects in the popular YouTube VFX Artists React series by Corridor Crew, in which the crew "react to" and "fix" the media affect of different VFX-intensive movies.[8]

Transmission of media affects

I have an affinity with disintegration effects. I remember from my early childhood trying to look at one thing for too long, and inevitable reaching a threshold at which that thing would visually break down and perception deteriorate. This was a game I used to play as a child, playing with attention and distraction, mutating myself into a state of trance or autohypnosis, absorbed, diverted, making myself nebulous. Through early experimentation with my own eyes as a visualization technology, within childhood's world of the chaos of sensation, I sensed (or discovered) the disconnected nature of reality. This particular technoperceptual habituation might be behind my enduring attunement to simulacra and its disintegrative possibilities. The demons of *Charmed* are encountered via spell, metabolized, and then disintegrated. They become ephemeral phenomena, which accord with demonological accounts of them as fundamentally mobile creatures.

But perhaps I like *Charmed* demons mainly because of my preference for *past techniques of making species visible*, the business of bestiaries. In popular contemporary culture, the demon is an organism from hell, out of history (discontinuous with us). They are uncivilized incarnations of a threatening proximity not of this world. And who knows demons best today? The technical animators, working in VFX Industry, department of creature design. Computer technical animation is an undisciplinary microworld, situated in transnational commercial production for mass culture, where *hacker skills are transduced to sensitized transmedia knowledge as they pass from the plane of heuristic techno-methodology to an interpretive plane of composing visual sense or "appeal"*. To think of the space of a CG software, I am using Martha Kenney's definition of microworld, a space where protocols and equipments are standardized to facilitate the emergence and stabilization of new objects.[9]

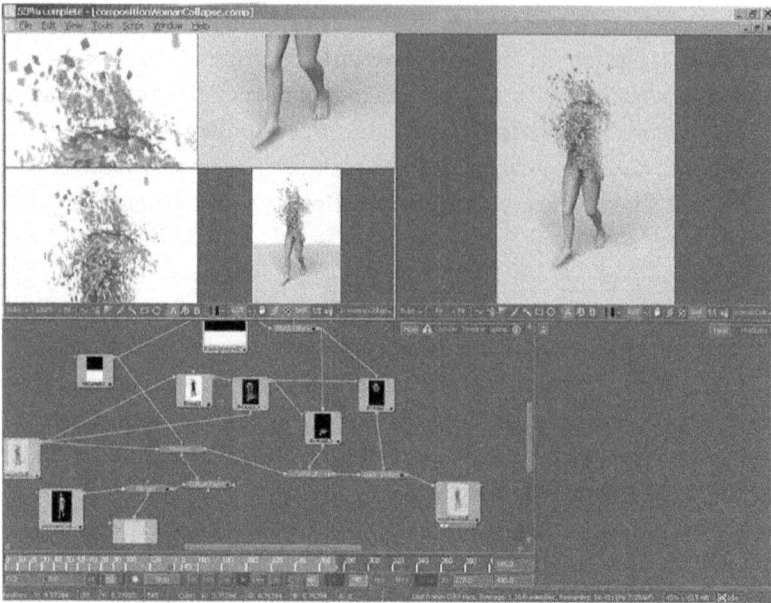

A disintegrative body rendered in Maya and composed in Fusion (eyeon). Being part of the technical animation industry, over the years I built and rebuilt collapsing bodies and disintegrative rigs for mesh objects. Sina Seifee, 2007

To get close to a lived texture of nonhuman nonanimal creatureliness, the technical animators have to sense the complexity of synthetic life through modeling (wealth of detail) and rigging (enacting structure). In other words, they need to get skilled at using digital phenomena (calculative abstraction) to create affectively positive encounters (appeal) with analogue body subjects that are irreducible to discrete mathematical states (the audience). This is a form of "open skill",[10] a context-contingent tactically oriented form of understanding or responsiveness. Creature animation defined as such is, essentially, a hacker's talent.

Following this understanding of technical animation, I want to highlight one of its actual practices as the focal point of interest in this writing, namely *rigging*. Rigging can be understood as staging and controlling "movement" within a limited computational structure (the microworld). Rigging is the talent associated with bringing an environment into transformational particularities using itself. It involves movement between the code space of the software environment (structural determination) and techniques they generate in response to that environment (emergent practice). In other words, the givens of computer graphics software are continually reworked in the creative responses CG hackers develop in relation to the microworld with which they interact. Rigging understood as such, is a workaround practice that both traverses and exceeds the stratified data of its microworld.

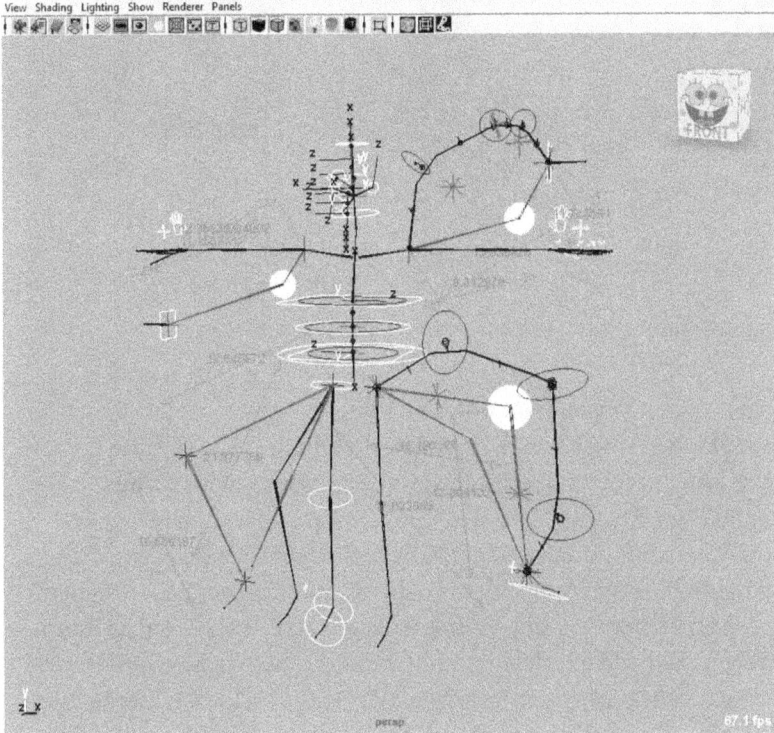

RANDOM / Maya Advance Rigging by Blender Sushi 2012, using Maya, underlying skeleton with IK/FK switch, muscle spline, spline IK, knee lock, and deformable head, https://mayaspiral.blogspot.com

Rigging almost always involves making a quality of liveliness through movement. That means, technical animators, through designing so-called rigs, have to create an *envelopment*: a complex form of difference between the *analogue* (somatic bodily techniques as the source of perceiving movement) and the *digital* (analytical ways of conceptualizing movement). This envelopment (skin) reduces what is taken as a model to codified tendencies that encourage and prohibit specific forms of movement and action. As such, rigging is a technological site where bodies are dreamed up, reiterated, or developed.

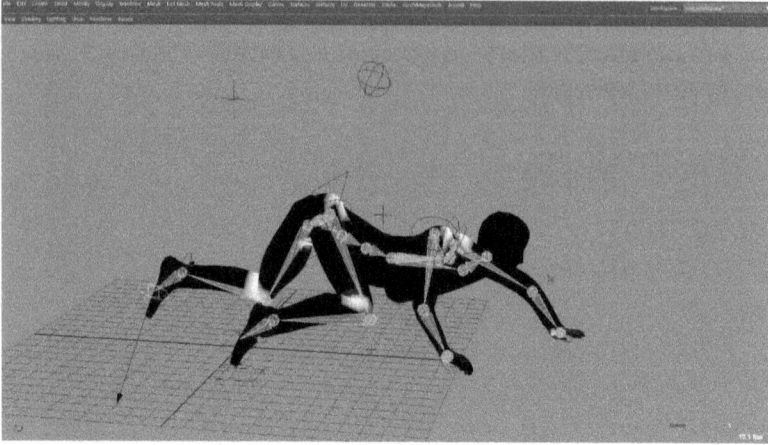

A simple rigged bipedal character in Maya. Sina Seifee, 2020

Animal animation industry

In his research on the nature of skill in computer multiplayer games, James Ash suggests that the design of successful video games depends on creating "affective feedback loops between player and game".[11] This is a quality of elusivity in the game's environment and its mode of interaction with the players, which is predicated on management and control of contingency itself. This is achieved by interactively testing the relation between the code space (game) and the somatic space (users). Drawing on Ash's insights, I would like to ask how affective quality of liveliness are distributed in the assemblages of various human and technical actors that make up rigging? Exploding demons; what kind of animal geography is that? This is a question of a non-living multi-species social subject in a technically mediated world. I follow Eben Kirksey's indication of the notion of *species* as a still useful "sense-making tool"[12] and propose that the demon's disintegrative body is a form of grasping species with technologies of visualization. In this case, rigging is part of the imagined species that is grasped through enacting (*disintegrativity* as its morphological characteristics).

Enacting is part of the material practice of learning and unlearning what is to be something else. To enact is to express, to collect and compose a part of the reality that needs to be realized and affirmed by the affects. To (re)enact something is a mutated desire to construct the invisible and mobile forces of that thing. Enactment is not just "making", it is part of much larger fantasy practices and realities. The most obvious examples are religion and marketing as two institutions that depend on the enactments of fans (of God or the brand). The new-media fandom (collectivities of fans) venture in a social and collaborative engagement with corporate engineered products. But as Henry Jenkins has argued, this engagement is highly ambiguous.[13] Technical animators often behave like fans of their own cultural milieu. For instance when the Los Angeles based visual effects company Corridor Crew tells their story of fixing the bad visual effects of the *Star Wars* franchise, they enact a fan-culture by modifying and thus creating a variation. They participate in shaping a techno-cognitive context for engagement with *Star Wars* that operates the same story (uniform cultural memory) but has an intensity of its own (potential for mutation).[14] As we can see in the case of Corridor Crew, technical animation is always a materially heterogeneous work. The animators don't sit on their desks, they enact all sorts of materialities. Animators use somatic intelligibility (embodiment) to fuse with their tools and become visual meaning-making machines that mutually embody their creatures. Therefore, the disintegration rig can be thought of as a human-machine enactment of a mixed-up species, a makeshift assemblage of human-demon-machinic agency enacting morphological transformations—bringing demon species into being. Doing demons is a social practice.

The animation industry is a complex set of talents and competencies associated with the distribution and transmission of media affects. Within VFX-intensive storytelling as one of the fastest growing markets of our time,[15] animation designers work to create artifacts potent with positively affective responses. The ways in which affect can be manipulated or preempted is a complex and problematic process.[16] Industrial model of distributed production is coalescence of conflicting agencies, infrastructures, responsibilities, skills, and pleasures where none of them is fully in command.[17] Animation technologies has evolved alongside the mass entertainment techno-capital market as a semi-disciplinary apparatus and its constituent player: fans, hackers, software developers, corporates, and pirate kingdoms. I prefer to use the term "hacker" (disorganized workaround practices) when referring to the talents of technical animators. CG hackers working in each other's hacks and rigs, through feedbacked assemblages of skill sharing, tutorial videos, screenshots, scripts, help files, shortcuts. The assemblages are made of layers of codes and tools built on each other, nested folders in one's own computer, named categories by oneself and others, horde of text files and rendered test JPGs, and so on. These are (en-/de-)crypting extended bodies of subjectively constructed through the communal technological fold interpreted as the 3D computer program. An ecology of pragmatic workaround practices that Amit Rai terms "collective practices of habituation", which Katie King might call "distributed embodiments, cognitions, and infrastructures at play".

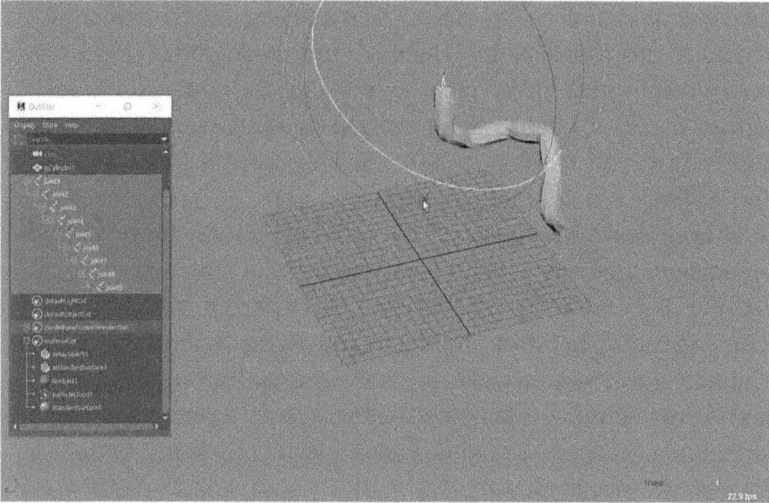

FK (forward kinematics) simple one-dimensional rigging in Maya, Sina Seifee, 2020. The rotation value of each "joint" is accumulated through the chain

I propose to understand CG hackers and technical artists with their practices of habituation, as craft-intensive. This implies understanding them as intimately connected with a particular microworld, the knowledge of which comes through skilled embodied practice that subsist over longer periods of time. I worked for some time as a generalist technical animator for both television and cinema, many years ago. An artisan's life and a set of skills that I acquired in my youth, which are still part of my repertoire of know-hows that makes me expressive today. As many others have argued,[18] crafters attune to their materials, becoming subject to the processes they are involved in. Then, rigging as a skill can be understood as a form of pre-conceptual practice. By pre-conceptual I mean what Benjamin Alberti refers

42

to as processes through which concepts find their way into actualities. Skilled practice is also the mark of the maker's openness to alterity.[19] An alterity in relation to that which the machinic entity becomes quasi-other or quasi-world.[20] Is it possible to invoke epistemological intimacy (a way of grasping one's own practice) through the processes of crafts? What is *Charmed*'s answer to this?

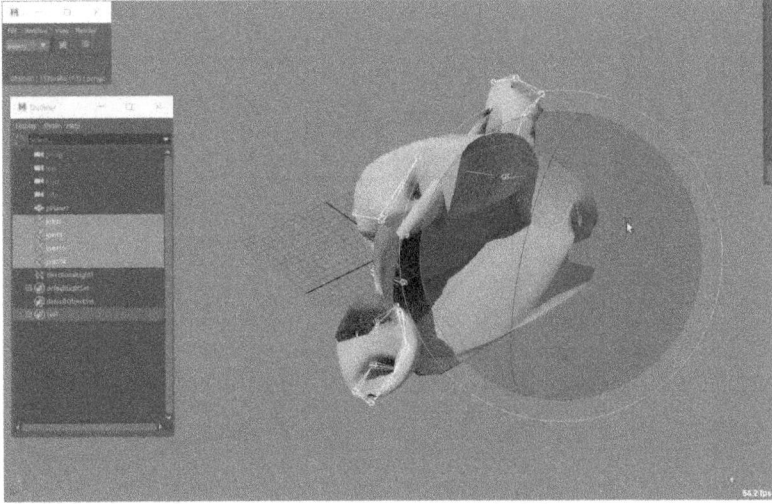

FK one dimensional rigging in Maya. Sina Seifee, 2020

Demon disintegration zoomorphic writing technology

CG stands for computer graphics, but also for many more things, *computational gesture*, and *creature generator*. In the example of demon disintegration that I gave earlier, I suggested the presence of zoomorphic figures (demons) as an indication for thinking about rigging as a bundle of the digital (calculative abstraction), the analogue (body appeal), and the nonhuman (zoomorphic physiology). Zoomorphic figures are historically bound with animation technologies. The design and rigging of "creatures" are part of every visual effects training program and infused in the job description. Disney Animation Studios is the example of critical and commercial success through mastery over anthropomorphized machines. Animation has been a technology of zoomorphic writing.

Automata and calligraphy's mimetic figures

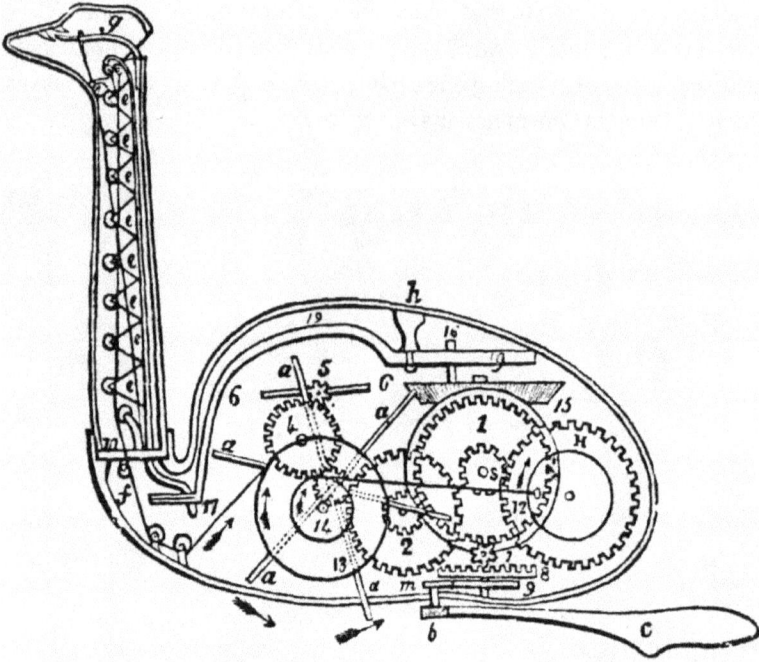

Engraving of *Digesting Duck*, an automaton in the form of a duck, created by Jacques de Vaucanson, 1739. Image from *A Dictionary of Arts, Manufactures, and Mines*, 1839

Zoomorphic writing technologies are not new. The clockwork animals, those attendant mammalian attachments, were bits of kinematic programming able to produce working simulacrums of a living organism. Perhaps rigging is a manifestation of the desire to produce and study automata. For Golem, that unfortunate unformed limb, the rig was YHWH, the name of the God. Another witness is a variation of calligraphy, the belle-lettre style of enfolding animals into letters, which is as old as writing itself. The particular volumetric regime of making animal shapes with calligraphy operates by confusing pictorial and lexical attributes, mobilizing a sort of wit in order to animate imaginary and real movements. Mixing textuality and figurality is something like a childhood experience. A kind of word-puzzle which uses figurative pictures with alphabetical shapes. It is a game of telescoping language through form, schematizing a space where the animal's body and language form one gestalt. In my childhood I was indeed put into a calligraphy course, which I eventually opted out of.

Although extremely short, my calligraphy training taught me how the world passes through the mechanized, technical, and skillful pressure of the pen, hand, color, paper, and eye as an assemblage. At that time I experienced calligraphy as an entirely uncharismatic technology. Yet I found myself spending endless hours making mimetic figures with writing. I felt how making animals with calligraphy conflates language and image and thus makes it liable to move in many unpredictable directions. The power of the latent, the hidden relationships, the interpretable. A state of multistability that I enjoyed immensely as a child.

Rigging demons as an occasion of contemporary zoomorphic writing technology suggests that the enfoldment of "morph" (the transformation of an image by computer) and "zoon" (nonhuman animals) is both that which nonhumans shape and that which gives shape to nonhumans. Bodies of demons in the software are enveloped with the appropriate rig for a specific transmodification (movement, disintegration, etc). But because of the presence of zoomorphism—like the case of calligraphy—they don't move as pure presuppositions. In rigging the deformation and movement are always in question.

Zoomorphic writing, opaque lapis-lazuli based paint and gold on paper, Iran, 12th century

Rigging as prosthetic technology

Following an understanding of technical animation habits in terms of their descriptive capacities, or *a pre-conceptual craft-intensive zoomorphic writing practice*, I would like to enlarge the understanding of rigging as an essentially prosthetic technology. Prosthetics simply means the extended body. They are vivid illustrations of human-technology relations in terms of the body (prosthetics are perhaps the exact opposite of Morton's hyperobjects). As the philosopher of virtual embodiment, Don Ihde has argued that the extended body signifies itself through the technical mediation. In this sense the body of the technical animator is an extended lived-body, a machine-infused neuro-physical body. Benefiting from a notion of apparatus developed by Karen Barad, namely apparatus understood as a sort of specific physical argument (fixed parts establishing a frame of reference for specifying "position"),[21] rigging can be thought of as a sort of articulation. We can now ask how rigging, as a specific prosthetic embodiment of the technologically enhanced visualization apparatus, matters to practices of knowing about the world, species, and demons?

Manual understanding abstract animals

As I have been showing, technical animators are *manual understanders* of nonhuman cyber-physiology. They have to be good at two things: morphology and its mathematization, or to be more precise, analytic geometry. Analytic geometry is not necessarily Euclidean or rigid body dynamics, because it also covers curved spaces, n-dimensional spaces, volumetric space, phase space, etc. As I was being self-educated in 3D animation, I learned to understand the space of the software as a n-dimensional manifold; X, Y, Z, the dimension of time, of texture, of audio, and so on. The particular way that technical animators look at nonhumans (animal or nonanimal) creates a mode of abstraction that reduces the state of amorphousness (model) to position and structure, like an anatomy, or as I call it, a rig. Less concerned with external resemblance (shading), rigging is particularly busy with building internal homologies. It is a comprehensible order (skeleton) that permits systematic animation, but also allows complexities and accidents to occur.

Homology is a morphological correspondence primarily determined by relative positions and connections. As soon as technical animators start thinking about rigging, they are doing anatomical work, a science of form. They use comparative biological intuition to imagine an isomorphic system of relations. Through building an abstract animal, they respond to the question of morphological correspondence or analogue. They become thinkers of organic folding. Analogue in homological terms means when a part or organ in one assemblage (an imagined animal) is isomorphic (it has the same function) to another part or organ in a different assemblage (virtual microworld). Rig is the analogue of the animal's body.

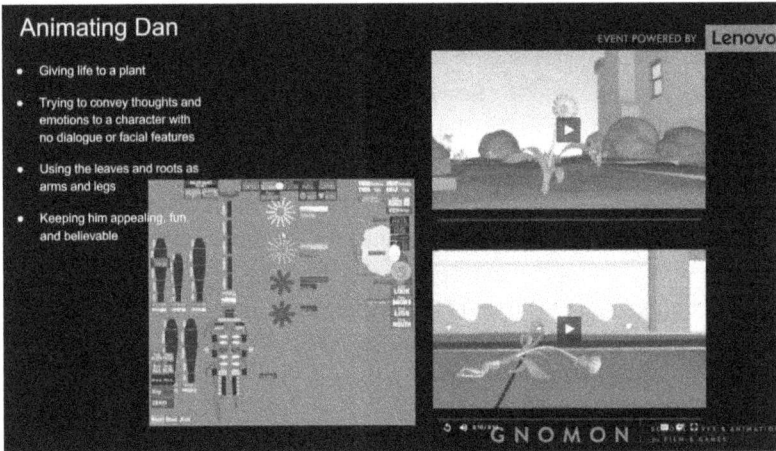

In their presentation of the project hosted by The Gnomon Workshop, *Weeds: The Making of an Animated Short Film*, a group of Disney tech-artists working on a distributed project that they did on their personal time, talk about how they cared for the dandelion in the process of rigging Dan in 3D animation *Weeds*.[22] Kevin Hudson, one of the animators, mentions how he started with attention and observation (opening their bodies to a variety of affective states): "The inspiration for the story came when I was out front of my house pulling weeds that pop up in my lawn. I looked across my driveway at my neighbor's yard, which was never watered, and the lawn is dead with only a few dying dandelions clinging to the edge of the sidewalk".[23] In the talk, we can see how the creation of "appeal" is understood as the creation of "care" in the animation culture industry. In the making of Dan, the pictorial effect of appeal is done to Dan's face as the substance of subjective singularity. Faciality as the medium of the anthropomorphic expression of the facial body (for example in *Weeds* the whole body becomes an expressive face) is one of the main mediums of the animation industry. The artists of Disney draw from understandings of the mammalian-affective structure (face) and technical agencies (rig) to create interactive dramas of psychological bonding.

My prosthetic experience with CG affirms with Ihde's notion of multistability. Technologies are multistable. That means they have unpredictable side-effects and are embeddable in different ways, in different cultures.[24] In a world where technologies and humans interactively constitute one another, I find Ihde's variational method-ology quite useful. It simply means that through variations, and not only through epistemic breakdowns, new gestalts can be forefronted. Fan based contents are generated precisely by variational creativity in the multistable plane of consumption. Ihde's variational approach is to be understood in contrast to the epistemological breakdown as a revelatory means of knowing—when something that had usually been taken for granted, under breakdown conditions, gets revealed in a new way. Following Ihde's indication, we can think of mechanisms of the production of differences as variations (how something varies, and is not simply breaking down) in the routines of rigging. They are technologies that are both effective and failing, obscuring and mak-ing visible the nonhumans that hackers like to realize. Through abstract speculation and variational (craft-intensive) inspection of the mundane technological mediation of monsters, I have been trying to propose a case for the heterogeneous relationships between human beings, the world and for artifacts used for mediation. I have been doing that to think about this question: How do CG hackers *make their animals more real?* In order to extend my response to that ques-tion, and still taking myself as an empirical access point, I will look at my extended being at work with computer graphics and make a case for phantom limbs.

Mastery of the phantom limb

I like to propose that prosthetic skills are intimately connected to the mastery of the phantom limb. Phantom limb is a technique of cogni-tive prosthesis, which allows for the creation of artificial limbs. A post-amputation phenomenon, phantom limbs are the sensation of missing limbs. Elizabeth Grosz has discussed the problematic and uncontainable status of the body in biology and psychology, and that the phantasmatically lost limbs are persistently part of our hermeneutic-cultural body. Is the embodiment through technologies, the technoperceptual habituation of the 3D software, a mode of engagement with the *body image?* Over longer periods of time, medi-ating technology can become an artificial limb for the subject. It can reach a state of instrumental transparency. That means that through skilled embodied practices the technical animator's interaction with their microwork achieves an intuitive character, a techno-perceptual bodily self-experience. The n-dimensional space of the animation software becomes part of the condition of one's access to spatiality. It becomes one's "body image". Simply put, the *body image* is the picture of our own body which we form in our mind. It is experienced viscerally and is always anatomically fictive and distorted. The concept of *body image*, coined by psychoanalyst Paul Schilder and neurologist Henry Head is a schema (spatiotemporally structured model) that mediates between the subject's position and its environment.

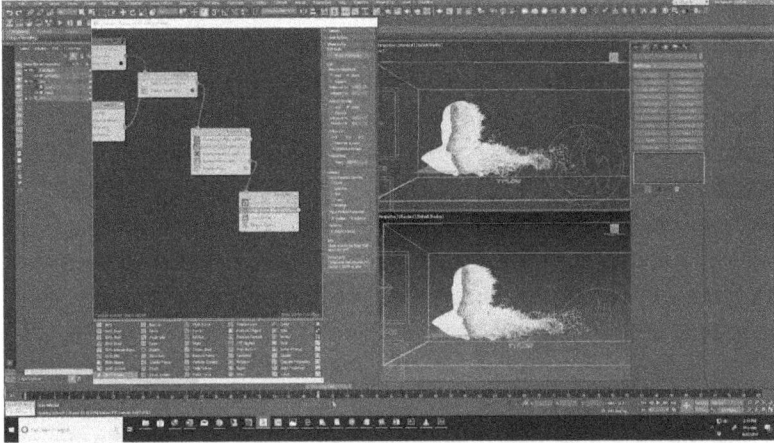

Allan McKay's tutorial on doing disintegration effects.[25] McKay is known for the dissemination visual effects that he achieved as the digital artist for the movie *Blade: Trinity*, 2004. Increased over the years, perhaps tripped by the 2018 film *Avengers: Infinity War*, a whole family of disintegration effects have become part of the entertainment industry's volumetrics[26]

A strange experience of engagement with phantom limbs can be found in religion. In Catholic theology to be sanctified involves the ritual of mortification of the flesh. Mortification refers to an act by which an individual or group seeks to put their sinful parts to death. As both an internal and external process, mortification involves exactly the continuity of missing parts (of the soul) with the living parts. Lacan called it "imaginary anatomy" and designated it as part of the genesis of the ego. Grosz makes note of this and further gives the example of a child becoming a subject through the development of its *body image*, in various libidinal intensities. Sensations are projected onto the world, the world's vicissitudes are introjected back into the child. The child's *body image* gets gradually constructed and invested in stages of libidinal development: The oral stage and the mouth, the anal stage and the anus, and so on. Children's bodies, like the process of modeling, move from a state of amorphousness to a state of increasing differentiation.[27]

Actors learn to constantly use the concept of *body image*. In an acting group that I was part of in the early 2000s, part of our training was to control and distort the *body image* at will in order to insinuate real affective states in one's self. Without naming it as such, we learned how the *body image* can shrink and expand. How it can give body parts to the outside world and can incorporate external objects. This is a mode of engagement with the phantom limb, in which the subject stimulates a state of possession of the body through external means. This is also the case in music improvisation. Everyone who has improvised with a musical instrument knows that playing music is not merely a technical problem of tool-use. I have been playing *setar* on and off for 20 years. Setar is a string-based instrument, and like lute it is played with the index finger. I learned it through a tacit and cognitive apprenticeship (not using notation), starting when I was still a teenager. Mastering a musical instrument as such becomes something personal, distributive, and bodily contextual. The strange phenomena of "mood" in playing the setar—which is the key to its mastery—is perhaps part of the difficulty of learning how to play the instrument. Getting into the mood is precisely the libidinal problem of how the instrument becomes psychically invested, how it becomes a cathected part of the *body image*.

Rigging as the mastery of the phantom limb made sense to my young self. As a shy teenager I was experiencing a discord between my psychical, idealized self-image (*body image*) and my actual unde-sired lived-body that felt like a biological imposition. As Grosz has also mentioned, teenagehood is precisely the age for philosophical desire to transcend corporeality and its urges. My relationship with CG technologies can be understood through ambivalent responses within puberty to the threat of inconsistency of the world. I was changing my *body image* through visualization of phantom limbs. And thus escaping a state of dispossession (a state of freedom from phan-toms). This is what I am calling *counter dispossession through engage-ment with the phantom limb*. A mode of prosthetic cognitive engage-ment with phantom limbs, perhaps against what Descartes warned as the deception of the inner senses. I am still attached to the world of unbelievable images, with its own immanent forms of movement. Witches exploding the body schema of the demons.

Demonological intimacy

What I am proposing here is to make a site of negotiation with the cyberbox of CG spaces, and to recognize rigging as a mode of engagement with such spaces. Rigging is a trajectory-enhancing device, another trajectory of human-nonhuman relational being that happens in the digital interface. If we take CG animation with its often nonhuman-referenced starting-point, and its prosthetic phenomenology as an extended technologically mediated nurturer of zoomorphic bodies, we can ask the following questions. Which species are socialized through machinic agency of rigging practices? What is the body schema of the hacker in CG as a microworld where there is no near or far? What is experienced as their Gestalt? What kind of grasp is automatically localized? What are their phantom limbs? These are all questions of volumetric regimes. In this essay I have been trying to create a site where responses to these inquiries can be constructed and played with, by observing myself playing and giving a bit more specificity to the demons of *Charmed*. And taking the hints that Grosz and Ihde give, understand myself as to be thinking and acting in the midst of the pervasive proliferation of technoperceptual phantom limbs.

To think of demon vanquishing visual effects as a model of synthesis, implies learning to see old and new forms of confusion, attachment, subjectivity, agency, and embodiment in mass media technoculture. *A postmodern machinic fantasy in which animators are technical computational de-amputators, exploding the guts of demons.* This is a supra-reality hybrid craft in digital form that suggests a mode of intimacy with nonhumans ambivalence. In demon rigging technical animation, the demon arrives as an older model of agency to inspire causality. It is a computer-cyberspace machinic intimacy but also demonological. Demonology is not necessarily only an ecclesiastic discourse (related to the church), but a variational practice of empirically verifying hybrid human-animal creatures from long-standing popular conceptions of a shared non-fictive reality. Call it a fandom spin-off of theology. They are part of the vast repertoire of composite and cross-disciplinary network of nonhuman causality and transmedia writing (bestiary).

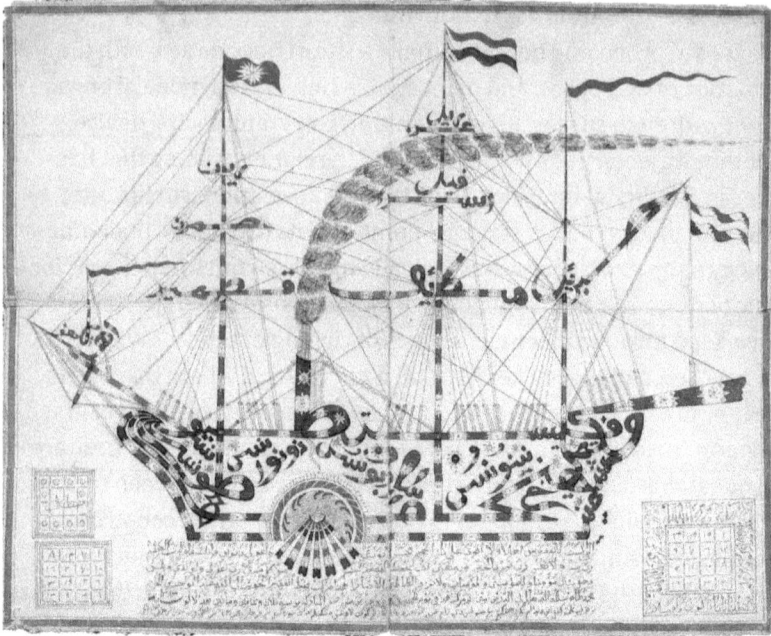

Talisman in the form of a warship (with the names of the "Seven Sleepers of Ephesus") signed by Abdul Wahid ibn al-Haji Muhammad Tahir, Indonesia, 1866, Bodleian Library, University of Oxford. Coined as a technique of sailing vessels, *rigging* is not a metaphoric thought. It refers rather to a cheat, a hacker's talent, in which one selects and puts components in place to allow them to function in a particular way

In order to make a scene (not an argument) about computerized zoopoetics, and learn something new about the perceptual selectivity of the CG hackers tangled in social machinery of animation tools, I tried to attend to my technohabitual experiences as a CG generalist amidst an increasing awareness of the multistable nature of media technologies. This was done by patterning of scales: the scale of individual attention to particular fringes of one's own mini experiences, and the scale of the experience of a shared inhabited world. I couldn't help using "we" (and "our") more than once in the essay. The determiner "we" is a simple magic spell, a transcendental metaphysical charm through which one speaker becomes many. I associated myself with the "we", to evoke the possibility of a witnessable scenographic truth-telling, in order to *demonstrate* (to vanquish and to fabricate simultaneously) a multidimensional microworld of effective rigging in CG, where the social conjoiner of *we* would matter. Did I evoke *Charmed* and *Corridor Crew* as part of this "we"? And, is "we" a sympoiesis or an acknowledgment of a true collective difference? Is "we" always needed to pull back to include alternate knowledge worlds? Like how it is done in soap operas.

Perhaps my relationship with *Charmed* is like Prue, Phoebe and Piper to their demons, between love and vanquish. I have been using the notion of multistability to think about the relationships that bind humans to virtual explosive demons as their significant "other" (according to *Charmed*). In *Rigging Demons*, a digital folktale, I have proposed rigging as a sensory medium (a mode of nearness and appropriation) and as exosomatic practice (prosthetic): extending part of one's subjectivity beyond the skin through engagements with digital animation technologies as phantom limbs. Every demonic dematerialization in *Charmed*, every vanquishment, is also a relinquishing—of materializing forces that create a network out of that which this essay is inspired. This text is itself part of the play with the consciousness of technical animator, CG interface, soap opera, my affective involvement (being spellbound to the series), and an unmetabolized speciation in the style of bestiaries. Exploding demons area visceral non-mammalian animality located within a spacetime that is coordinated by commercial entertainment, transmedia writing technologies, zoosemiotic registers, and all sorts of agents that I am part of. I have been trying to propose a variational understanding of the 3D software as an interactive and augmented microworld of objects, beings, zoons and tools for the visualization of multistable cognitions, a form of transnational knowledge work that many agents (market, demons, machines, hackers) are involved in but none is in full control over.

Notes

1. ↑ Katie King, "A Naturalcultural Collection of Affections: Transdisciplinary Stories of Transmedia Ecologies Learning." *The Scholar & Feminist Online* 10, no.3 (2012), http://sfonline.barnard.edu/feminist-media-theory/a-naturalcultural-collection-of-affections-transdisciplinary-stories-of-transmedia-ecologies-learning/0/.

2. ↑ Isabelle Stengers, "Introductory Notes on an Ecology of Practices," *Cultural Studies Review* 11, no. 1 (August 2013), https://doi.org/10.5130/csr.v11i1.3459.

3. ↑ Gregory Bateson, *Steps to an Ecology of Mind. Collected Essays in Anthropology, Psychiatry, Evolution, and Epistemology* (University of Chicago Press, 1972).

4. ↑ Amit S. Rai, *Jugaad Time: Ecologies of Everyday Hacking in India* (Durham: Duke University Press, 2019).

5. ↑ Patricia T. Clough, "In the Aporia of Ontology and Epistemology: Toward a Politics of Measure." *The Scholar &*

Feminist Online 10, no. 3 (2013).

6. ↑ Lorraine J. Daston and Katharine Park, "Unnatural Conceptions: The Study of Monsters in Sixteenth- and Seventeenth-Century France and England." *Past & Present 92* (August 1981): 20-54.

7. ↑ Ernest Mathijs, "Referential acting and the ensemble cast." *Screen 52*, no. 1 (March 2011): 89–96, https://doi.org/10.1093/screen/hjq063.

8. ↑ Corridor Crew, "We Fixed the Worst VFX Movie Ever," published 2020, https://youtu.be/MYKrnNedhOw.

9. ↑ Martha Kenney, *Fables of Attention: Wonder in Feminist Theory and Scientific Practice*, UC Santa Cruz (2013), https://escholarship.org/uc/item/14q7k1jz

10. ↑ John Sutton, "Batting, Habit and Memory: The Embodied Mind and the Nature of Skill", *Sport in Society - Cultures, Commerce, Media, Politics* 10, no. 5 (August 2007): 763-786. https://doi.org/10.1080/17430430701442462.

11. ↑ James Ash, "Architectures of Affect: Anticipating and Manipulating the Event in Processes of Videogame Design and Testing," *Environment and Planning D: Society and Space* 28 (4): 653-671. (2010), https://doi.org/10.1068%2Fd9309

12. ↑ Eben Kirksey, "Species: a praxiographic study," *Journal of the Royal Anthropological Institute* 5, no. 21 (October 2015): 758-780. https://doi.org/10.1111/1467-9655.12286.

13. ↑ Henry Jenkins, *Convergence Culture - Where Old and New Media Collide* (New York: NYU Press, 2006).

14. ↑ Corridor Crew, "We Made Star Wars R-Rated," published 2019, https://youtu.be/GZ8mwFiXlP8

15. ↑ Rama Venkatasawmy, "The Evolution of VFX-Intensive Filmmaking in 20th Century Hollywood Cinema: An Historical Overview," *TMC Academic Journal* 6, no. 2 (2012), 17-31.

16. ↑ James Ash, "Architectures of Affect: Anticipating and Manipulating the Event in Processes of Videogame Design and Testing," *Environment and Planning D: Society and Space* 28, no. 4 2010, 653-671. https://doi.org/10.1068%2Fd9309.

17. ↑ King, "A Naturalcultural Collection of Affections"; Anna Tsing and Elizabeth Pollman, "Global Futures: The Game," in *Histories of the Future* (Durham: Duke University Press, 2005). https://doi.org/10.1215/9780822386810-005; Jenkins, *Convergence Culture*, 2006.

18. ↑ Benjamin Alberti, "Art, craft, and the ontology of archaeological things," *Interdisciplinary Science Reviews* 43, no. 3-4 (December 2018), 280-294, https://doi.org/10.1080/03080188.2018.1533299; Don Ihde, *Bodies in Technology* (Durham: University of Minnesota Press, 2002); Ash, "Architectures of Affect"; Richard Sennett, *The Craftsman* (New York: Yale University, 2009).

19. ↑ Alberti, "Art, craft, and the ontology of archaeological things".

20. ↑ Ihde, *Bodies in Technology*.

21. ↑ Karen Barad, "Posthumanist Performativity: Toward an Understanding of How Matter Comes to Matter," *Signs* 28, no. 3 (2003): 801-831, https://doi.org/10.1086/345321.

22. ↑ Kevin Hudson et al, "Animating Dan," *The Gnonom Workshop*, Facebook, posted November 8, 2017, https://www.facebook.com/thegnomonworkshop/videos/10155383708888037017

23. ↑ "Weeds (2017 film)," Wikipedia, accessed October 21, 2021, https://en.wikipedia.org/wiki/Weeds_(2017_film)#Conception_and_writing

24. ↑ Don Ihde, "Forty Years in the Wilderness," in *Postphenomenology: A Critical Companion to Ihde* (Albany: SUNY Press, 2006).

25. ↑ © Allan McKay, "Thanos 3DS Max Particles Thanos VFX Tutorial (tyFlow & Phoenix FD)," accessed October 21, 2021, https://youtu.be/OHOM8QpeysU.

26. ↑ "Thanos Disintegration" search results on YouTube, https://www.youtube.com/results?search_query=Thanos+Disintegration.

27. ↑ Elizabeth Grosz, *Volatile Bodies: Toward a corporeal feminism* (Bloomington: Indiana University Press, 1994).

Dis-orientation and its Aftermath

Jara Rocha, Femke Snelting

We remain physically upright not through the mechanism of the skeleton or even through the nervous regulation of muscular tone, but because we are caught up in a world.[1]

This text is based on three items selected from *The Possible Bodies Inventory*. We settled for inventorying as a method because we want to give an account of the structural formations conditioning the various cultural artifacts that co-compose 3D polygon "bodies" through scanning, tracking and modeling. With the help of the multi-scalar and collective practice of inventorying, we attempt to think along the agency of these items, hopefully widening their possibilities rather than pre-designing ways of doing that too easily could crystallize into ways of being. Rather than rarefying the items, as would happen through the practice of collecting, or pinning them down, as in the practice of cartography, or rigidly stabilizing them, as might be a risk through the practice of archiving, inventorying is about continuous updates, and keeping items available.

Among all of the apparatuses of the Modern Project that persistently operate on present world orderings, naming and account-giving, we chose the inventory with a critical awareness of its etymological origin. It is remarkably colonial and persistently productivist: inventory is linked to invention, and thereby to discovery and acquisition.[2] The culture of inventorying remits us to the material origins of commercial and industrial capitalism, and connects it with the contemporary database-based cosmology of techno-colonialist turbo-capitalism. But we've learned about the potentials embedded in Modern apparatuses of designation and occupation, and how they can be put to use as long as they are carefully unfolded to allow for active problematization and situated understanding.[3] In the case of Possible Bodies, it means to keep questioning how artifacts co-habit and co-compose with techno-scientific practices, historically sustained through diverse axes of inequality. We urgently need research practices that go through axes of diversity.

The temporalities for inventorying are discontinuous, and its modes of existence pragmatic: it is about finding ways to collectively specify and take stock, to prepare for eventual replacement, repair or replenishment. Inventorying is a hands-on practice of readying for further use, not one of account-giving for the sake of legitimization. As an "onto-epistemological" practice,[4] it is as much about recognizing what is there (ontological) as it is about trying to understand (epistemological). Additionally, with its roots in the culture of manufacture, inventorying counts on cultural reflection as well as on action. This is how inventorying as a method links to what we call "disobedient action-research", it invokes and invites further remediation that can go from the academic paper to the bug report, from the narrative to the diagrammatic, and from tool mis-use to interface re-design to the dance-floor. It provides us with inscriptions, de-scriptions and re-interpretations of a vocabulary that is developing all along.

For this text, we followed the invitation of Sara Ahmed, "to think how queer politics might involve disorientation, without legislating disorientation as a politics".[5] We inventoried three items, *Worldsettings for beginners*, *No Ground* and *Loops*, each related to the politics of dis-orientation. In their own way, these artifacts relate to a world that is becoming oblique, where inside and outside, up and down switch places and where new perspectives become available. The items speak of the mutual constitution of technology and bodies, of matter and semiotics, of nature and culture and how orientation is managed in tools across the technological matrix of representation. The three items allow us to look at tools that represent, track and model "bodies" through diverse cultural means of abstraction, and to convoke their aftermath.

Item 007: Worldsettings for beginners

Author(s) of the item: **Blender community**
Year: **1995**
Entry date: **March 2017**
Cluster(s) the item belongs to: **Dis-orientation**

If the point of origin changes, the world moves but the body doesn't.[6]

In computer graphics and other geometry-related data processing, calculations are based on Cartesian coordinates, that consist of three different dimensional axes: x y and z. In 3D-modelling, this is also referred to as "the world". The point of origin literally figures as the beginning of the local or global computational context that a 3D object functions in.

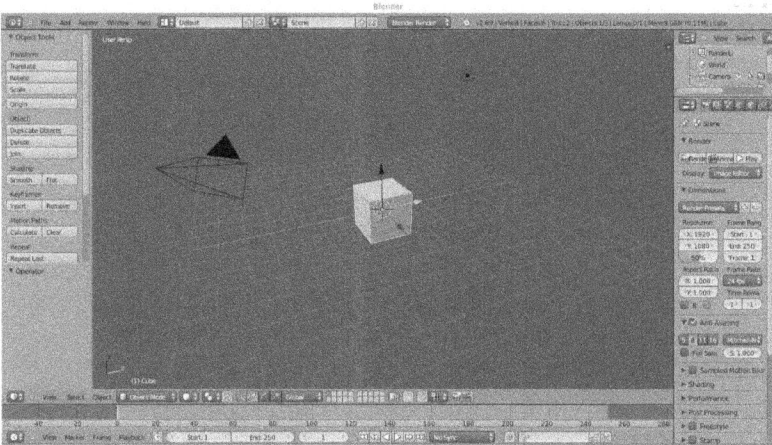

Screenshot Blender 2.69, 2017

Using software manuals as probes into computational realities, we traced the concept of "world" in Blender, a powerful Free, Libre and Open Source 3D creation suite. We tried to experience its process of "worlding" by staying on the cusp of "entering" into the software. Keeping a balance between comprehension and confusion, we used the sense of dis-orientation that shifting understandings of the word "world" created, to gauge what happens when such a heady term is lifted from colloquial language to be re-normalized and re-naturalized in software. In the nauseating semiotic context of 3D modeling, the word "world" starts to function in another, equally real but abstract space. Through the design of interfaces, the development of software, the writing of manuals and the production of instructional videos, this space is inhabited, used, named, projected and carefully built by its day-to-day users.

In Blender, virtual space is referred to in many ways: the mesh, coordinate system, geometry and finally, "the world". Each case denotes a constellation of x, y, z vectors that start from a mathematical point of origin, arbitrarily located in relation to a 3D object and automatically starting from x = 0, y = 0, z = 0. Wherever this point is placed, all other planes, vertices and faces become relative to it and organize around it; the point performs as an "origin" for subsequent trans-formations.

In the coordinate system of linear perspective, the vanishing point produces an illusion of horizon and horizontality, to be perceived by a monocular spectator that marks the center of perception and reproduction. Points of origin do not make such claims of visual stability.

> *The origin does not have to be located in the center of the geometry (e.g. mesh). This means that an object can have its origin located on one end of the mesh or even completely outside the mesh.*[7]

There is not just one world in software like Blender. On the contrary, each object has its own point of origin, defining its own local coordinates. These multiple world-declarations are a practical solution for the problem of locally transforming single objects that are placed in a global coordinate system. It allows you to manipulate rotations and translations on a local level and then outsource the positioning to the software that will calculate them in relation to the global coordinates. The multi-perspectives in Blender are possible because in computational reality, "bodies" and objects exist in their own regime of truth that is formulated according to a mathematical standard. Following the same processual logic, the concept of "context" in Blender is a mathematical construct, calculated around the world's origin. Naturalized means of orientation such as verticality and gravity are effects, applied at the moment of rendering.

> *Blender is a two-handed program. You need both hands to operate it. This is most obvious when navigating in the 3D View. When you navigate, you are changing your view of the world; you are not changing the world.*[8]

The point of origin is where control is literally located. The two-handedness of the representational system indicates a possibility to shift from *navigation* (vanishing point) into *creation* (point of origin), using the same coordinate system. The double agency produced by this ability to alternate is only tempered by the fact that it is not possible to take both positions at the same time.

> *Each object has an origin point. The location of this point determines where the object is located in 3D space. When an object is selected, a small circle appears, denoting the origin point. The location of the origin point is important when translating, rotating or scaling an object. See Pivot Points for more.*[9]

The second form of control placed at the origin is the 3D manipulator that handles the rotation, translation, and scaling of the object. In this way, the points of origin function as pivots that the worlds are moved around.

An altogether different cluster of world metaphors is at work in the "world tab". Firmly re-orienting the virtual back in the direction of the physical, these settings influence how an object is rendered and made to look "natural".

> *The world environment can emit light, ranging from a single solid color, physical sky model, to arbitrary textures.*[10]

The tab contains settings for adding effects such as mist, stars, and shadows but also "ambient occlusion". The Blender manual explains this as a "trick that is not physically accurate", maybe suggesting that the other settings are. The "world tab" leaves behind all potentials of multiplicity that became available through the computational understanding of "world". The world of worlds becomes, there, impossible.

Why not the world? On one hand, the transposition of the word "world" into Blender functions as a way to imagine a radicaly interconnected multiplicity, and opens up the possibility for political fictions derived from practices such as scaling, displacing, de-centering and/or alternating. On the other hand, through its linkage to (a vocabulary) of control, its world-view stays close to that of actual world domination. Blender operates with two modes of "world". One that is accepting of the otherness of the computational object, somehow awkwardly interfacing with it, and another that is about restoring order, back to what is supposedly real. The first mode opens up to a widening of the possible, the second prefers to stick to the plausible, and the probable.

Item 012: No Ground

Author(s) of the item: **mojoDallas, Hito Steyerl**
Year: **2008, 2012**
Entry date: **5 March 2017**
Cluster(s) the item belongs to: **Dis-orientation**

*A fall toward objects without reservation, embracing a
world of forces and matter, which lacks any original stability and sparks the sudden shock of the open: a freedom
that is terrifying, utterly deterritorializing, and always
already unknown. Falling means ruin and demise as well
as love and abandon, passion and surrender, decline and
catastrophe. Falling is corruption as well as liberation, a
condition that turns people into things and vice versa. It
takes place in an opening we could endure or enjoy,
embrace or suffer, or simply accept as reality.*[11]

This item follows Hito Steyerl in her reflection on disorientation and
the condition of falling, and drag it all the way to the analysis of an
animation generated from a motion capture file. The motion capture
of a person jumping, is included in the Carnegie-Mellon University
Graphics Lab Human Motion Library.[12] Motion capture systems,
including the one at Carnegie Mellon, typically do not record information about context, and the orientation of the movement is made relative to an arbitrary point of origin.[13]

In the animated example, the position of the figure in relation to
the floor is "wrong", the body seems to float a few centimeters above
ground. The software relies on perceptual automatisms and plots a
naturalistic shadow, taking the un-grounded position of the figure
automatically into account: if there is a body, a shadow must be computed for. Automatic naturalization: technology operates with material diligence. What emerges is not the image of the body, but the body
of the image: "The image itself has a body, both expressed by it's construction and material composition, and [...] this body may be inani-
mate, and material".[14]

No ground is an attempt to think through issues with situatedness that appear when encountering computed and computational
bodies. Does location work at all, if there is no ground? Is displacement a movement, if there is no place? How are surfaces behaving
around this no-land's man, and what forces affect them?

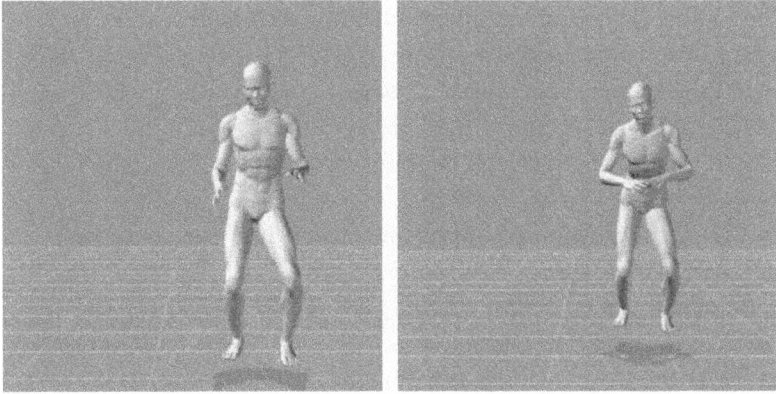

Animation: mojoDallas, © Mike Sutton, 2008, https://www.youtube.com/watch?v=ZakpoLqXhyI

The found-on-the-go ethics and path dependence that conditions the computational materialities of bodies worry us. It all appears too imposing, too normative in the humanist sense, too essentialist even. What body compositions share a horizontal base, what entities have the gift of behaving vertically? How do other trajectories affect our semiotic-material conditions of possibility, and hence the very politics that bodies happen to co-compose? How can these perceptual automatism be de-clutched from a long history of domination, of the terrestrial and extraterrestrial wild that is now sneaking into virtual spheres?[15]

We suspect that this is due to a twist in the hierarchy between gravitational forces. This twistdoes not lead to collapse but results in a hallucinatory construction of reality, filled with floating "bodies". If we want to continue using the notions of "context" and "situation" for the cultural analysis of so-called bodies that populate the pharmaco-pornographic, military and gamer industries and their imaginations; to attend to their immediate political implications, we need to reshape our understanding of them. It might be necessary to let go of the need for "ground" as a defining element for the very existence of the "body", though this makes us wonder about the agencies at work in this un-grounded embodiments. If the land is for those who work it, then who is working the ground?[16]

Disorientation involves failed orientations: bodies inhabit spaces that do not extend their shape, or use objects that do not extend their reach.[17]

The co-constitution of so-called bodies and technologies shatters all dream of stability, the co-composition of foreground and background crashes all dreams of perspective. When standing just does not happen due to a lack of context or a lack of ground, even if it is a virtual one, the notion of standpoint does not work anymore. Situation, though, deserves a second thought.

The political landscape of turning people into things and vice-versa recalls the rupture of "knowing subjects" and "known objects" that Haraway called for after reading the epistemic use of "standpoint" in Harding,[18] which asked for a recognition of the "view from below" of the subjugated: "to see from below is neither easily learned nor unproblematic, even if "we" "naturally" inhabit the great underground terrain of subjugated knowledges".[19] The emancipatory romanticism of Harding does not work in these virtual renderings. The semiotic-material conditions of possibility that unfold from Steyerl's above description are conditions without point, either when viewed from standing or from below.

What would be the implication of displacing our operations based on unconsolidated matter that in its looseness asks for eventual anchors of interdependence? How could we transmute the notion of situatedness in order to understand the semiotic-material conditionings of 3D rendered bodies that affect us socially and culturally through multiple managerial worldings?

Here the "body" is neither static nor falling: it is floating. Here we find that Haraway's "situatedness" does not match when we try to manage potential vocabularies for the complex forms of worldmaking and its embodiments in the virtual. What can we learn from the conditions of floating that is brought to us by the virtual transduction of the Modern perspective, in order to draft an account-giving apparatus of present presences? How can that account-giving be intersectional with regards to the agencies implied, respectful of the dimensionality of time and aging, and responsible with a political history of groundness?

Floating is the endurance of falling. It seems that in a computed environment, falling is always in some way a floating. There is no ground to fall towards that limits the time of falling, nor is the trajectory of the fall directed by gravity. The trajectory of a floating or persistently falling body is always already unknown.

In the dynamic imagination of the 3D-animation, the ground does not exist before movement is generated, it only appears as an afterthought. Everything seems upside down: the foundation of the figure is deduced, does not pre-exist its movement. Is there actually no foundation, or does it just mean that it appears in every other loop of movement? Without the ground, the represented body can be seen as becoming smaller and if so, that would open the question on dimensionality and scaleability. But being surface-dependent, it is received as moving backwards and forwards: the Modern eye reads one shape that changes places within a territory. Closer, further, higher, lower: the body arranges itself in perspective, but we must attend the differences inherent in that active positioning. The fact that we are dealing with an animatedmoving body implies that the dimension of time is brought into the conversation. Displacement is temporary, with a huge variation in the gradient of time from momentary to persistent.

In most virtual embodiments, the absolute tyranny of the conditions of gravity do not operate. In a physical situation (a situation organized around atoms), falling on verticality is a key trajectory of displacement; falling cannot happen horizontally upon or over stable surfaces. For the fleshy experienced, falling counts on gravity as a force. Falling seems to relate to liquidity or weightlessness, and grounding to solidity and settlement of matters. Heaviness, having weight, is a characteristic of being-in-the-world, or more precisely: of being-on-earth, magnetically enforced. Falling is depending on gravity, but it is also – as Steyerl explains – a state of being un-fixed, ungrounded, not as a result of groundbreaking but as an ontological lack of soil, of base. Un-fixed from the ground, or from its representation.[20]

Nevertheless, when gravity is computed, it becomes a visual-representational problem, not an absolute one. In the animation, the figure is fixed and sustained by mathematical points of origin – but to the spectator from earth, the body seems unfixed from its "natural soil". Hence, in a computational space, other directions become possible thanks to a flipped order of orientation: the upside-down regime is expanded by others like left-right, North-South and all the diagonal and multi-vortex combinations of them. This difference in space-time opens up the potential for denaturalized movements.

Does falling change when the conditions of verticality, movement and gravity change? Does it depend on a specific axis of these conditions? Is it a motion-based phenomenon, or a static one? Is it a rebellion against the force of gravity, since falling here functions in a mathematical rather than in a magnetic paradigm? And if so, "who" is the agent of that rebellion?

At minute 01:05, we find a moment where two realities are juxtaposed. For a second, the toe of the figure trespasses the border of its assigned surface, glitching a way out of its position in the world, and bringing with it an idea of a pierceable surface to exist upon... opening it up for an eventual common world.

In the example, the "feet" of the figure do not touch the "ground". It reminds us that the position of this figure is the result of computation. It hints at how there are rebellious computational semiotic-material conditions of possibility at work. We call them semiotic because they are written, codified, inscribed and formulated (alphanumerically, to start with). We call them material since they imply an ordering, a composition of the world, a structuring of its shapes and behaviors. Both conditions affect the formulation of a "body" by considering weight, height and distance. They also affect the physicality of computing: processes that generate computing's pulses in electromagnetic circuits, power network use, server load, etc.

When the computational grid is placed under the feet of the jumping figure, materialities have to be computed, generated and located "back" and "down" into a "world". Only in relation to a fixed point of origin and after having declared its world to make it exist, the surrounding surfaces can be settled. Accuracy would depend on how those elements are placed in relation to the positioned "body". But accuracy is a relational practice: body and ground are computed separately, each within their own regime of precision. When the rendering of the movement makes them dependent on the placement of the ground, their related accuracy will appear as strong or weak, and this intensity will define the kind of presence emerging.

Thinking present presences can not rely on the lie of laying. A thought about agency can neither rely on the ground to fall towards nor on the roots of grass that emerge from it. How can we then invoke a politics of floating not on the surface but within, not cornered but around and not over but beyond, in a collective but not as a grass-roots movement? Constitutive conditioning of objects and subjects is absolutely relational, and hence we must think of and operate with their consistencies in a radically relational way: not as autonomous entities but as interdependent worldings. Ground and feet, land and movement, verticality and time, situatedness and axes: the more of them we take into account when giving account of the spheres we share, the more degrees of freedom we are going to endow our deterritorialized and reterritorialized lives with.

The body is a political fiction, one that is alive; but a fiction is not a lie.[21] And so are up, down, outside, base, East and South and presence.[22] Nevertheless, we must unfold the insights from knowing how those fictions are built to better understand their radical affection on the composition of what we understand as "living", whether that daily experience is mediated, fleshly or virtually.

Item 022: Loops

Author(s) of the item: **Golan Levin, Merce Cunningham, OpenEnded group, Buckminster Fuller**
Year: **2009, 2008, 1971, 1946**
Entry date: **November 2016**
Cluster(s) the item belongs to: **Dis-orientation**

Loops entered the Possible Bodies inventory for the first time through an experiment by Golan Levin.[23] Using an imaging technique called Isosurfacing, common in medical data-visualization and cartography, Levin rendered a motion recording of Cunningham's performance *Loops*. The source code of the project is published on his website as golan_loops.zip. The archive contains, among c-code and several Open Framework libraries, two motion capture files formatted in the popular Biovision Hierarchy file format, rwrist.bvh.txt and lwrist.bvh.txt. There is no license included in the archives.[24]

Following the standard lay-out of .bvh, each of the files starts with a detailed skeleton hierarchy where in this case, WRIST is declared as ROOT. Cascading down into carpals and phalanges, Rindex is followed by Rmiddle, Rpinky, RRing and finally Rthumb. After the hierarchy section, there is a MOTION section that includes a long row of numbers.

Just before he died in 2009, Merce Cunningham released the choreography for *Loops* under a Creative Commons Attribution-Non-commercial-Share Alike 3.0 license. No dance-notations were published, nor has The Merce Cunningham Trust included the piece in the collection of *68 Dance Capsules* that provides "an array of assets essential to the study and reconstruction of this iconic artist's choreographic work".[25]

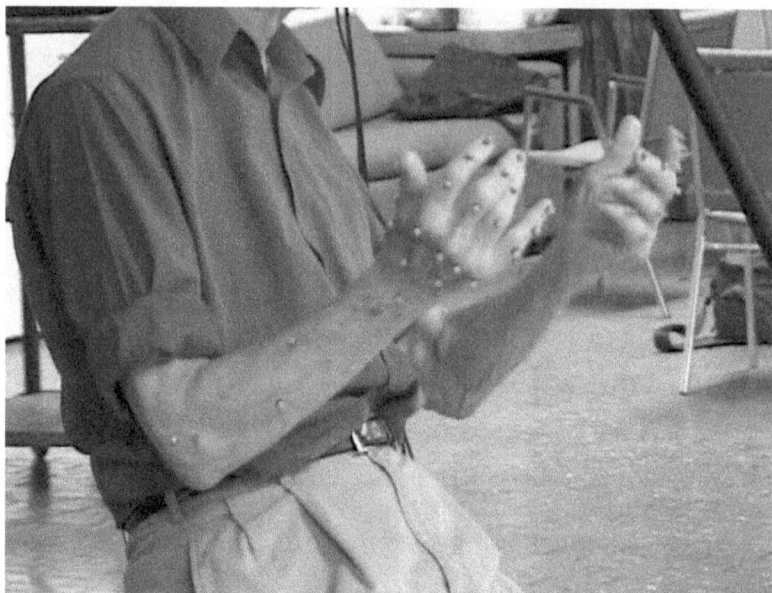

Merce Cunningham and OpenEnded group, *Loops: Take 1* (hand-held), 2001

From the late nineties, the digital art collective OpenEnded group worked closely with Cunningham. In 2001, they recorded four takes of Cunningham performing *Loops*, translating the movement of his hands and fingers into a set of datapoints. The idea was to "Open up Cunningham's choreography of Loops completely" as a way to test the idea that the preservation of a performance could count as a form of distribution.[26]

The release of the recorded data consists of four compressed folders. Each of the folders contains a .fbx (Filmbox) file, a proprietary file format for motion recording owned by software company Autodesk, and two Hierarchical Translation-Rotation files, a less common motion capture storage format. The export files in the first take is called Loops1_export.fbx and the two motion capture files loops1_all_right.htr and loops1_all_left.htr. Each take is documented on video, one with hand-held camera and one on tripod. There is no license included in the archives.

In 2008, the OpenEnded group wrote custom software to create a screen based work called *Loops*. Loops runs in real time, continually drawing from the recorded data. "Unique? — No and yes: no, the underlying code may be duplicated exactly at any time (and not just in theory but in practice, since we've released it as open source); yes, in that no playback of the code is ever the same, so that what you glimpse on the screen now you will never see again".[27] The digital artwork is released under a GPL v.3 license, but seeing interpretations of *Loops* made by other digital artists such as Golan Levin, OpenEnded group declared that they did not have any further interest in anyone else interpreting the recordings: "I found the whole thing insulting, if not to us, certainly to Merce".[28]

Cunningham developed *Loops* as a performance to be exclusively executed by himself. He continued to dance the piece throughout his life in various forms until arthritis forced him to limit its execution to just his hands and fingers.[29]

In earlier iterations, Cunningham moved through different body parts and their variations one at a time and in any order: feet, head, trunk, legs, shoulders, fingers. The idea was to explore the maximum number of movement possibilities within the anatomical restrictions of each joint rotation. Stamatia Portanova writes: "Despite the attempt at performing as many simultaneous movements as possible (for example, of hands and feet together), the performance is conceived as a step-by-step actualization of the concept of a binary choice".[30]

A recording of *Loops* performed in 1975 is included in the New York Public Library Digital Collections, but can only viewed on site.[31] Cunningham danced *Loops* for the first time in the Museum of Modern Art in 1971. He situated the performance in front of *Map (Based on Buckminster Fuller's Dymaxion Airocean World)*, a painting by his friend Jasper Johns. Roger Copeland describes *Loops* as follows: "In much the same way that Fuller and Johns flatten out the earth with scrupulous objectivity, Cunningham danced in a rootless way that demonstrated no special preference for any one spot". Later on, in the same book, "Consistent with his determination to decentralize the space of performance, Cunningham's twitching fingers never seemed to point in any one direction or favor any particular part of the world represented by Johns's map painting immediately behind him".[32]

In one of the rare images that circulates of the 1971 performance, we see Cunningham with composer Gordon Mumma in the background. From the photograph it is not possible to detect if Cunningham is facing the painting while dancing *Loops*, and whether the audience was seeing the painting behind or in front of him.

Cunningham met Buckminster Fuller in 1948 at Black Mountain College. In an interview with Jeffrey Schnapp, he describes listening to one of Fuller's lectures: "In the beginning you thought, this is absolutely wonderful, but of course it won't work. But then, if you listened, you thought, well maybe it could. He didn't stop, so in the end I always felt like I had a wonderful experience about possibilities, whether they ever came about or not".[33]

With *The Dymaxion Airocean World Map*, Buckminster Fuller wanted to visualize planet earth with greater accuracy. In this way "humans will be better equipped to address challenges as we face our common future aboard Spaceship Earth". The description of the map on the Buckminister Fuller Institute website is followed by a statement that "the word Dymaxion, Spaceship Earth and the Fuller Projection Map are trademarks of the Buckminster Fuller Institute. All rights reserved".[34]

The Dymaxion Airocean Projection divides the surface of the earth into 20 equilateral spherical triangles in order to produce a two-dimensional projection of the globe. Fuller patented the Dymaxion map at the US Patent Office in 1946.[35]

Aftermath

The inventorying of the *items 007, 012* and *022* has allowed us to think through three cultural artifacts with very different scales, densities, medias and durations. The items were selected because they align with a fundamental inquiry into 3D-infused imaginations of the "body" and their consequences, emerging through a set of questions related to orientation and dis-orientation. Additionally, the items represent the transdisciplinarity of the issues with 3D scanning, modeling and tracking, that touch upon performance analysis, math, cartography, intellectual property law and software studies.

In *Item 007: Worldsettings for beginners*, we explored the singular way in which the Cartesian coordinate system inhabits the digital by producing worlds in 3D modeling software, including the world of the so-called body itself. In *Item 012: No Ground*, we asked how situatedness can be meaningful when there is no ground to stand on. We wondered about which tools we might need to develop in order to organize forms, shapes and ultimately life when floating upon virtual disorientation. Finally in *Item 022: Loops*, we followed the embodiment of a choreographic practice, captured in files and legal documents; all the way up, and back to facing the earth.

The text provides evidence for some of the ways that inventorying could work as a research method, specifically when interrogating digital apparatuses and the ethico-political implications that are nested in the most legitimated and capitalized industries of techno-colonial totalizing innovations, in defining the limits of the fictional construction of fleshy matters: what computes as a body?

The main engine for Possible Bodies as a collective research, is to problematize the hegemonic pulsations in those technologies that deal with "bodies" in their volumetric dimension. In order to understand the (somato)political conditioning of our everyday, this research unfolds an intersectional practice, with a trans∗feminist sensibility through the prism of aesthetics and ethics.

Evidently, our questions both grew sharper and overflowed while studying the items and testing their limits, fueling Possible Bodies as a project. Inventorying opens up possibilities for an urgent mutation of that complex matrix, by diffracting from probabilistic normativity.

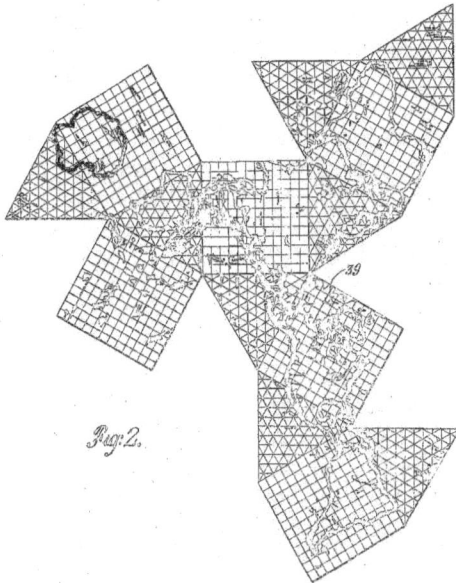

Buckminster Fuller, US Patent 2393676, Dymaxion Airocean Projection, 1946

Notes

1. ↑ Merleau-Ponty quoted in Sara Ahmed, *Queer Phenomenology: Orientations, Objects, Others* (Durham: Duke University Press, 2006).

2. ↑ From Medieval Latin "inventorium," alteration of Late Latin "inventarium;" "list of what is found," from Latin "inventus," past participle of "invenire," "to find, discover, ascertain". Online Etymology Dictionary, accessed April 21, 2021. http://www.etymonline.com/index.php?term=inventory.

3. ↑ Donna Haraway, "The promises of monsters: a regenerative politics for inappropriate/d others," eds. Lawrence Grossberg, Cary Nelson, and Paula A. Treichler, *Cultural Studies* (London: Routledge, 1992), 295-336.

4. ↑ Karen Barad, "Matter feels, converses, suffers, desires, yearns and remembers," in *New Materialism: Interviews & Cartographies,* eds. Rick Dolphijn, and Iris van der Tuin (London: Open Humanities Press, 2012).

5. ↑ Ahmed, *Queer Phenomenology. Orientations, Objects, Others.*

6. ↑ François Zajega in conversation with Possible Bodies, 2017.

7. ↑ "Individual Origins," *Blender Manual*, accessed April 10, 2021, https://docs.blender.org/manual/en/dev/editors/3dview/controls/pivot_point/individual_origins.html.

8. ↑ Gordon Fisher, *Blender 3D Basics Beginner's Guide* (Birmingham: Packt Publishing, 2014).

9. ↑ "Object Origin," *Blender Manual*, accessed April 10, 2021, https://docs.blender.org/manual/en/dev/scene_layout/object/origin.html.

10. ↑ "World," *Blender Manual*, accessed April 10, 2021, world.html https://docs.blender.org/manual/en/dev/render/eevee/world.html.

11. ↑ Hito Steyerl, "In Free Fall: A Thought Experiment on Vertical Perspective," *e-flux Journal #24* (April 2011), http://www.e-flux.com/journal/24/67860/in-free-fall-a-thought-experiment-on-vertical-perspective.

12. ↑ "CMU Graphics Lab Motion Capture Database," accessed April 10, 2021, http://mocap.cs.cmu.edu.

13. ↑ "Item 007: Worldsettings for beginners," *The Possible Bodies Inventory,* 2017.

14. ↑ Hito Steyerl, "Ripping reality: Blind spots and wrecked data in 3d," *European Institute for Progressive Cultural Policies,* (2017) accessed April 10, 2021, http://eipcp.net/e/projects/heterolingual/files/hitosteyerl/.

15. ↑ Donna Haraway, "The promises of monsters: a regenerative politics for inappropriate/d others."

16. ↑ *The Chiapas Media Project*, "Land Belongs to those Who Work It," 2005, https://vimeo.com/45615376.

17. ↑ Sara Ahmed, *Queer Phenomenology. Orientations, Objects, Others* (Duke University Press, 2006)

18. ↑ Sandra Harding, *The Science Question in Feminism* (Ithaca, NY: Cornell University Press 1986).

19. ↑ Donna Haraway, "Situated Knowledges: The Science Question in Feminism and the Privilege of Partial Perspective," *Feminist Studies,* 14 no. 3 (1998): 584.

20. ↑ Hito Steyerl, "In Free Fall: A Thought Experiment on Vertical Perspective", *e-flux Journal #24* (2011)

21. ↑ Paul B. Preciado, "Pharmaco-pornographic Politics: Towards a New Gender Ecology," *Parallax* vol. 14, no. 1 (2008): 105-117.

22. ↑ Jara Rocha, "Testing Texting South: a political fiction," in *Machine Research*, 2016.

23. ↑ "Item 024: Merce's Isosurface," *The Possible Bodies Inventory*, 2017.

24. ↑ On-line archives Golan Levin, accessed April 10, 2021. http://www.flong.com/storage/code/golan_loops.zip.

25. ↑ Larraine Nicholas and Geraldine Morris, *Rethinking Dance History: Issues and Methodologies* (Milton Park: Routledge, 2017).

26. ↑ This is precisely how the Merce Cunningham Dance Capsules website introduces itself, http://dancecapsules.merce.broadleafclients.com/index.cfm.

27. ↑ "Openended group", accessed April 10, 2021, http://openendedgroup.com.

28. ↑ Marc Downie and Paul Kaiser, "Drawing true lines", accessed April 10, 2021, http://openendedgroup.com/writings/drawingTrue.html.

29. ↑ Paul Kaiser quoted in Ashley Taylor, "Dancing in digital immortality. The evolution of Merce Cunningham's 'Loops'", *ScienceLine* (July 16, 2012), http://scienceline.org/2012/07/dancing-in-digital-immortality/

30. ↑ Stamatia Portanova, *Moving Without a Body* (Cambridge MA: MIT Press, 2012), 131.

31. ↑ "Changing steps [and] Loops, 1975-03-07", The New York Public Library Digital Collections, accessed April 10, 2021, https://digitalcollections.nypl.org/items/2103ccd0-e87e-0131-dc7f-3c075448cc4b.

32. ↑ Roger Copeland, *Merce Cunningham: The Modernizing of Modern Dance* (New York, Routledge, 2004), 247.

33. ↑ Jeffrey Schnapp, "Merce Cunningham: An Interview on R. Buckminster Fuller and Black Mountain College," August 31, 2016, https://jeffreyschnapp.com/2016/08/31/merce-cunningham-an-interview-on-r-buckminster-fuller-and-black-mountain-college.

34. ↑ "Dymaxion Map," Buckminister Fuller Institute, accessed April 10, 2021, https://www.bfi.org/about-fuller/big-ideas/dymaxion-world/dymaxion-map.

35. ↑ "Cartography: US2393676A," Google Patents, accessed April 10, 2021. https://www.google.com/patents/US2393676.

x, y, z (4 filmstills)

Jara Rocha, Femke Snelting

The volume of volumetric data that mining companies, hospitals, border agents and gaming industries acquire is ever increasing in scale and resolution. As a result, the usage of powerful software environments to analyse and navigate this digital matter grows exponentially. Imaging platforms draw expertise from computer vision, 3D-visualisation and algorithmic data-processing to join forces with Modern science. Obediently adhering to Euclidean perspective, they efficiently generate virtual volumes and perform exclusionary boundaries on the fly.

To interrogate the consequences of these alignments, x, y, z consists of four film stills from a movie-in-the making. The movie calls for queer rotations and disobedient trans*feminist angles that can go beyond the rigidness of axiomatic axes within the techno-ecologies of 3D tracking, modeling and scanning. It is an attempt to think along the agency of certain cultural artifacts, hopefully widening their possibilities beyond pre-designed ways of doing and being.

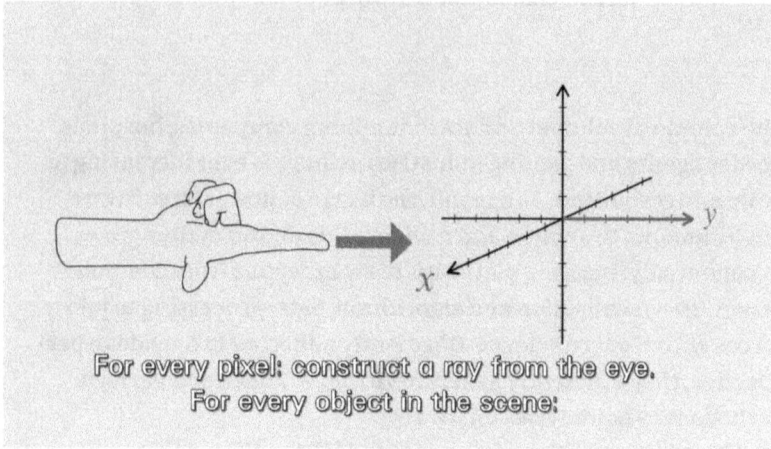

Item 014: *The Right-Hand Rule* + Item 105: *A ray from the eye*

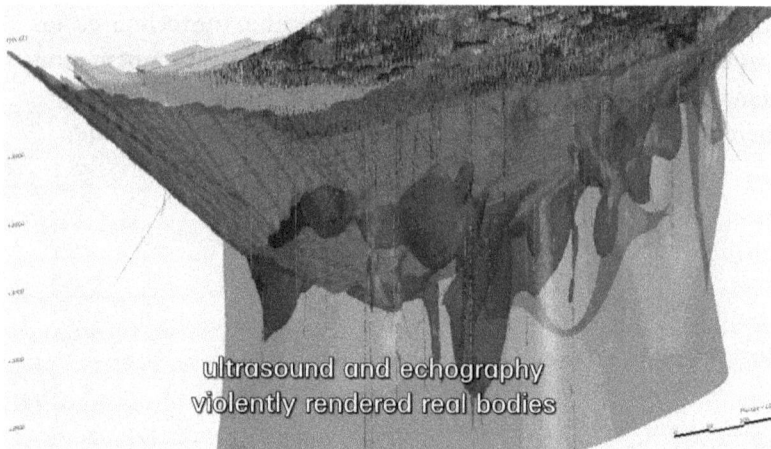

Item 090: *Model Our Planet* + Item 082: *Ultrasonic Dreams of Aclinical Renderings*

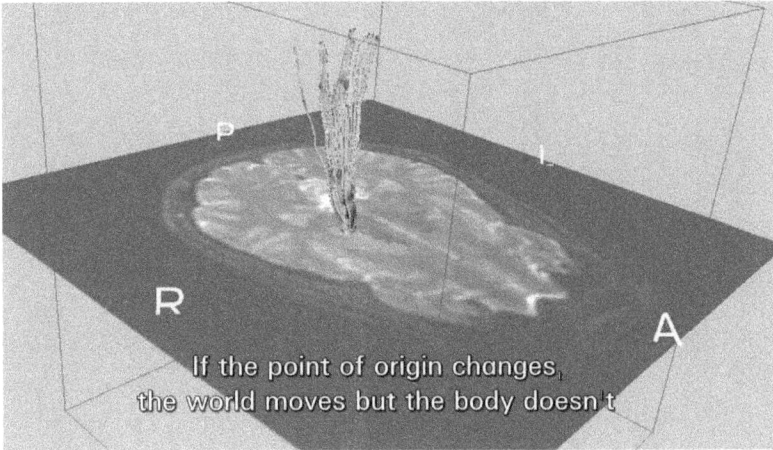

Item 098: *Region of interest* + Item 007: *Worldsetting for beginners*

Item 003: *Artist Drawing a Nude with Perspective Device* + Item 087: *The Crisis of Presence*

Invasive Imagination and its Agential Cuts

Jara Rocha, Femke Snelting

There is a conversation missing on the politics of computer tomography, on what is going on with data captured by MRI, PET and CT scanners, rendered as 3D-volumes and then managed, analyzed, visualized and navigated within complex software environments. By aligning medical evidence with computational power, biomedical imaging seems to operate at the forefront of technological advancement while remaining all too attached to Modern gestures of cutting, dividing and slicing. Computer tomography actively naturalizes Modern regimes such as Euclidean geometry, discretization, anatomy, ocularity and computational efficiency to create powerful political fictions: invasive imaginations and inventions that provoke the technocratic and scientific truth of so-called bodies. This text is a call for trans∗feminist[1] software prototyping, a persistent affirmation of the possibilities for radical experimentation, especially in the hypercomputational context of biomedical imaging.

1. Slice

In which we follow the emergence of a slice and its encounters with Euclidean geometry.

The appearance of the slice in biomedical imaging coincides with the desire to optimize the use of optical microscopes in the 18th century. Specimen were cut into thin translucent sections and mounted between glass, to maximize their accessible surface area and to be able to slide them more easily under the objective. Microtomography (after "tomos" which means *slice* in Greek), seems at first sight to be conceptually coherent with contemporary volumetric scanning techniques and computer tomography. But where microtomography produces visual access by physically cutting into specimen, computer tomography promises to stay on the outside. In order to affectively and effectively navigate matter, ocularity has been replaced by digital data-visualisation.

In computer tomography, "slice" stands for a data entity containing total density values acquired from a cross-section of a volume. MRI, PET or CT scanners rotate around matter conglomerates such as human bodies, crime scenes or rocks to continuously probe their consistency with the help of radiation.[2] The acquired data is digitally discrete but spatially and temporally ongoing. Only once turned into data, depths and densities can be cut into slices, and computationally flattened onto a succession of two-dimensional virtual surfaces that are back-projected so that each eventually resembles a contrasted black and white X-ray. Based on the digital cross-sections that are mathematically aligned into a stack, a third dimension can now be reverse-engineered. This volumetric operation blends data acquired at different micro-moments into a homogeneous volume. The computational process of translating matter density into numbers, reconstructing these as stacks of two-dimensional slices and then extrapolating additional planes to re-render three-dimensional volumes, is the basis of most volumetric imaging today.

Tomography emerged from a long-standing techno-scientific exploration that was fueled by the desire to making the invisible insides of bodies visible. It follows the tradition of anatomic experiments into a "new visual reality" produced by early x-ray imagery.[3] The slice was a collective invention by many: technologists, tools, users, uses, designers and others tied the increasing availability of computational capacity and to the mathematical theorem of an Austrian mathematician and the standardization of radio-densities.[4] Demonstrating the human and more-than-human entanglements of techno-scientific streams, the slice invoked multiple pre-established paradigms to provoke an unusual view on and inside of the world. Forty years later, most hospitals in the Global North have MRI and CT scanners operating around the clock.[5] In the mean time, the slice became involved in the production of multiple truths, as tomography was propagated along the industrial continuum of 3D: from human brain imaging to other influential fields of data-extraction such as mining, border-surveillance, mineralogy, large-scale fishing, entomology and archaeology.[6]

The acceleration produced by the probable jump from flattening x-rays to the third dimension can hardly be overestimated, a jump which is made even more useful because of the alleged "non-invasive" character of this technique: tomography promises visual access without the violence of dissection. Looking at the insides of a specimen which was traditionally conditioned by its death or by inducing *anaesthesia*, does not anymore require physical intervention.[7] But the persistence of the cross-cut, the fast assumptions that are made

about the non-temporality of the slice, the supposed indexical relation they have to matter, the way math is involved in the re-genera-
tion of densities and the location of tissues, all of it makes us wonder
about the not-non-invasiveness of the imagination at work in the
bio(info)technological tale. Looking is somehow always already an
operation.

Tomographic slices necessitate powerful software platforms to
be visualized, analyzed, rendered and navigated. We call such platforms "powerful" because of their extensive (and expensive) computational capacities, but also because of ways they embody authority
and truth-making. Software works hard to remove any trace of the
presence of the scanning apparatus and of the mattered bodies that
were once present inside of it. For slices to behave as a single volume
that is scanned at a single instant, they need to be normalized and
aligned to then neatly fit in the three orthogonal planes of X, Y and Z.
This automated process of "registration" draws expertise from computer vision, 3D-visualisation and algorithmic data-processing to
stack slices in probable ways.

From here on, the slices are aligned with the rigidity of Euclidean
geometry, a mathematical paradigm with its own system of truth, a
straight truth.[8] It relies on a set of axioms or postulates where the x, y
and z axes are always parallel, and where all corpo-real volumes are
located in the cubic reality of their square angles.[9] For reasons of efficiency, hardware optimization, path dependency and compatibility,
Euclidean geometry has become the un-questionable neutral spatial
norm in any software used for volumetric rendering, whether this is
gaming, flight planning or geodata processing. But in the case of biomedical imaging, x, y and z axes also conveniently fit the "sagittal",
"coronal" and "axial" planes that were established in anatomical science in the 19th century.[10] The slices have been made to fit the fiction
of medicine as seamlessly, as they have been made to fit the fiction of
computation.

Extrapolated along probable axis and obediently registered to the
Euclidean perspective, the slices are now ready to be rendered as
high-res three dimensional volumes. Two common practices from
across the industrial continuum of volumetric imaging are combined
for this operation: Ray-tracing and image segmentation. Ray-tracing
considers each pixel in each slice as the point of intersection with a
ray of light, as if it was projected from a simulated eye and then
encountered as a virtual object. "Imaging" enters the picture only at

the moment of rendering, when the ray-tracing algorithm re-inserts the re-assuring presences of both ocularity and a virtual internal sun. Ray-tracing is a form of algorithmic drawing which makes objects appear on the scene by projecting lines that originate from a single vantage point. It means that every time a volume is rendered, ray-tracing performs Dürer's enlightenment classic, *Artist drawing a nude with perspective device*.[11] Ray-tracing literally inverses the centralized god-like "vision" of the renaissance artist and turns it into an act of creation.

Image segmentation is the so-called non-invasive digital replacement of dissection. It starts at the boundaries rendered on each slice, assuming that a continuous light area surrounded by a darker one is indexical of the presence of coherent materiality; difference on the other hand signals a border between inside and outside. With the help of partially automatic edge detection algorithms, contrasted areas are demarcated and can subsequently be transformed into synthetic surfaces with the help of a computer graphics algorithm such as Marching Cubes. The resulting mesh- or polygon model can be rendered as continuous three dimensional volumes with unambiguous borders.[12] What is important here is that the doings and happenings of tomography literally *make* invisible insides visible.

From the very beginning of the tomographic process there has been an entanglement at work between computation and anatomy.[13] For a computer scientist, segmentation is a set of standard techniques used in the field of Computer Vision to algorithmically discern useful bits and pieces of images. When anatomist use the same term, they refer to the process of cutting one part of an organism from another. For radiologists, segmentation means visually discerning anatomical parts. In computer tomography, traditions of math, computation, perspective and anatomy join forces to perform exclusionary boundaries together, identifying tissue types at the level of single pixels. In the process, invisible insides become readable and eventually writable for further processing. Cut along all-too-probable sets of gestures, dependent on assumptions of medical truth, indexality and profit, slices have collaborated in the transformation of so-called bodies into stable, clearly demarcated volumes that can be operated upon. The making visible that tomography does is the result of a series of generative re-renderings that should be considered as operative themselves.[14] Tomography re-presents matter-conglomerates as continuous, stable entities and contributes strongly to the establishment of coherent materiality and humanness-as-individual-oneness. These picturings create powerful political fictions; imaginations and inventions that provoke the technocratic and scientific truth of so-called bodies.

The processual quantification of matter under such efficient regimes produces predictable outcomes, oriented by industrial concerns that are aligned with pre-established decisions on what counts as pathology or exploitation. What is at stake here is how probable sights of the no-longer-invisible are being framed. So, what implications would it have to let go of the probable, and to try some other ways of making invisible insides visible? What would be an intersectional operation that disobeys anthropo-euro-andro-capable projections? Or: how to otherwise reclaim the worlding of these possible insides?

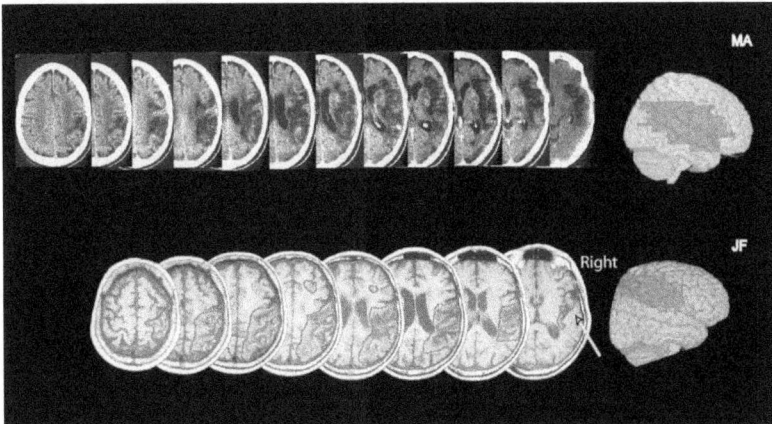

"Representative CT slices of patient MA (top row) and T1 MRI slices of patient JF (bottom row)" in Leon Y. Deouell, Diana Deutsch, Donatella Scabini, Nachum Soroker and Robert T. Knight, "No disillusions in auditory extinction: perceiving a melody comprised of unperceived notes", *Frontiers in Human Neuroscience* 2, no. 1: 15 (March 2008)

Albrecht Dürer, "Artist drawing a nude with perspective device", 1525

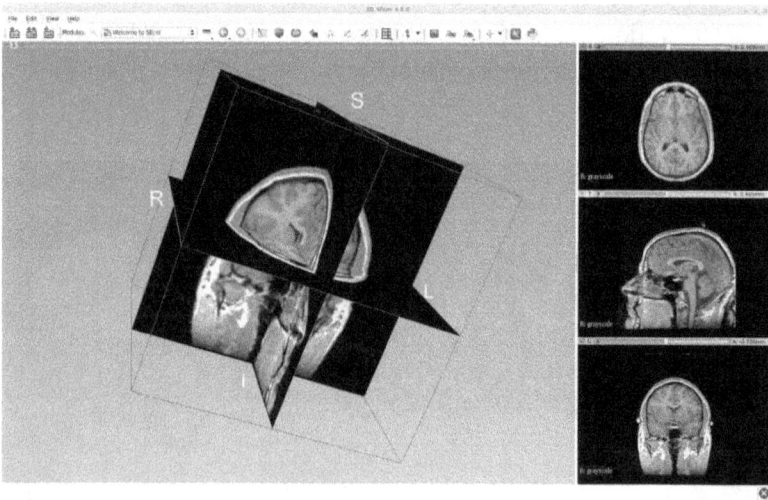

Basic image registration in Slicer v4.10.2 (screenshot)

2. Slicer

In which we meet Slicer, and its collision with trans∗feminist urgencies.

Feminist critical analysis of representation has been helpful in formulating a response to the kind of worlds that slices produce. But by persistently asking questions like: who sees, who is seen, and who is allowed to participate in the closed circuit of "seeing", such modes of critique too easily take the side of the individual subject. Moreover, it is clear that in the context of biomedical informatics, the issue of hegemonic modes of doing is more widely distributed than the problem of the (expert) eye, as will become increasingly clear when we meet our protagonist, the software platform Slicer. It is why we are interested in working through trans*feminist concepts such as entanglement and intra-action as a way to engage with the complicated more-than-oneness that these kinds of techno-ecologies evidently put in practice.

Slicer or or 3D-Slicer is an Open Source software platform for the analysis and visualization of medical images in research environments.[15] The platform is auto-framed by its name, an explicit choice to place the work of cutting or dividing in the center; an unapologetic celebration of the geometric norm of contemporary biomedical imaging. Naming a software "Slicer" imports the cut as a naturalized gesture, justifying it as an obvious need to prepare data for scientific objectivity. Figuring the software as "slicer" (like butcher, baker, or doctor) turns it into a performative device by which the violence of that cut is delegated to the software itself. By this delegation, the software puts itself at the service of fitting the already-cut slices to multiple paradigms of *straightness*, to relentlessly re-render them as visually accessible volumes.[16] In such an environment, any oblique, deviating, unfinished or *queer* cut become hard to imagine.

Slicer evolved in the fertile space between scientific research, biomedical imaging and the industry of scanning devices. It sits comfortably in the middle of a booming industry that attempts to seamlessly integrate hardware and software, flesh, bone, radiation, economy, data-processing with the management of it all. In the clinic, such software environments are running on expensive patented radiology hardware, sold by global technology companies such as Philips, Siemens and General Electric. In the high-end commercial context of biomedical imaging, Slicer is one of the few platforms that runs independent of specific devices and can be installed on generic laptops. The software is released under an Open Source license which invites different types of users to study, use, distribute and co-develop the project and its related practices. The project is maintained by a community of medical image computing researchers that take care of technical development, documentation, versioning, testing and the publication of a continuous stream of open access papers.[17]

At several locations in- and around Slicer, users are warned that this software is not intended for clinical use.[18] The reason Slicer positions itself so persistently outside the clinic might be a liability issue but seems most of all a way to assert itself as a prototyping environment meant to service the in-between of diagnostic practice and innovative marketable products.[19] The consortium managing Slicer draws in millions worth of US medical grants every year, already for more than a decade. Even so, Slicer's interface comes across as alarmingly amateurish, bloating the screen with a myriad of options and layers that are only vaguely reminiscent of the subdued sleekness of corresponding commercial packages. The all-over-the place impression of Slicer's interface coincides with its coherent mission to be a prototyping rather than an actual software platform. As a result, its architecture is skeletal and its substance consists almost entirely of extensions, each developed for very different types of biomedical research. Only some of this research concerns actual software development, most of it is aimed at developing algorithms for automating tasks such as anomaly detection or organ segmentation. The ideologies and hegemonies embedded in the components of this (also) collectively-developed-software are again confirmed by the recent adoption of a BSD license which is considered to be the most "business-friendly" Open Source license around.

The development of Slicer is interwoven with two almost simultaneous genealogies of acceleration in biomedical informatics. The first is linked to the influential environment of the Artificial Intelligence labs at MIT. In the late nineties, Slicer emerged there as a tool to demonstrate the potential of intervention planning. From the start, the platform connected the arts and manners of Quantitative Imaging to early experiments in robotic surgery. This origin story binds the non-clinical environment of Slicer tightly to the invasive gestures of the computer-assisted physician.[20]

A second, even more spectacular genealogy, is Slicer's shared history with the *Visible Human Project*. In the mid-nineties, as the volume of tomographic data was growing, the American Library of Science felt it necessary to publicly re-confirm the picturings with the visible insides of an actual human body, and to verify that the captured data responded to specifically mattered flesh. While the blurry black and white slices did seem to resemble anatomic structures, how to be sure that the results were actually correct?

A multi-billion dollar project was launched to materially re-enact the computational gesture of tomography onto actual flesh-and-blood bodies. The project started with the acquisition of two "volunteers", one convicted white middle-aged male murderer, allegedly seeking repentance through donating his body to science, and a white middle-aged female, donated by her husband. Their corpses where

first horizontally positioned and scanned, before being vertically stabilized in clear blue liquid, then frozen, and sewn into four pieces.[21]
Each piece was mounted under a camera, and photographed in a
zenithal plane before being scraped down 3 millimeter at a time, to be
photographed again. The resulting color photographs where digitized,
color-corrected, registered and re-rendered volumetrically in x, y, z
planes. Both datasets (the MRI-data and the digitized photographs)
where released semi-publicly. These two datasets, informally
renamed into "Adam" and "Eve" still circulate as default reference
material in biomedical imaging, among others in current versions of
Slicer.[22] Names affect matter; or better said: naming is always already
mattering.[23]

The mediatized process of the Visible Human Project coincided
with a big push for accessible imagining software platforms that
would offer fly-through 3D anatomical atlases, re-inserting Modern
regimes at the intersection of computer science, biomedical science
and general education.[24] The process produced the need for the
development of automatic registration and segmentation algorithms
such as the Insight Segmentation and Registration Toolkit (ITK), an
algorithm that is at the basis of Slicer.[25]

Slicer opens a small window onto the complex and hypercomputational world of biomedical imaging and the way software creates
the matter-cultural conditions of possibility that render so-called
bodies volumetrically present. It tells stories of interlocking regimes
of power which discipline the body; its modes and representations are
in a top-to-bottom mode. It shows how these regimes operate
through a distributed and naturalized assumption of efficiency that
hegemonically reproduces bodies as singular entities that need to be
clear and *ready* in order to be "healed". But even when we are critical
of how Slicer orders both technological innovation and biovalue as an
economy,[26] its licensing and positioning also create collective conditions for an affirmative cultural critique of software artifacts. We suspect that a F/LOSS environment responsibilizes its community to
make sure boundaries do not sit still. Without wanting to suggest that
F/LOSS itself produces the conditions for non-hegemonic imaginations, its persistent commitment to transformation is key for radical
experiments, and for trans*feminist software prototyping.

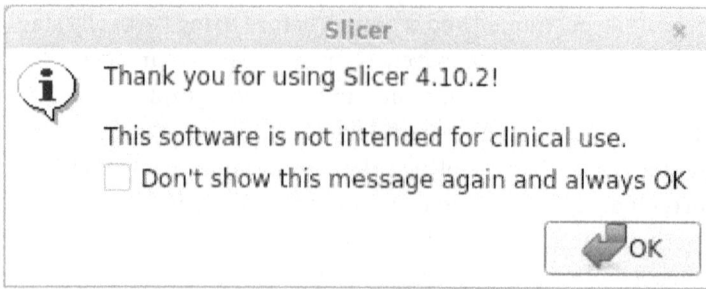

"Not for clinical use", Slicer v4.10.2 (screenshot)

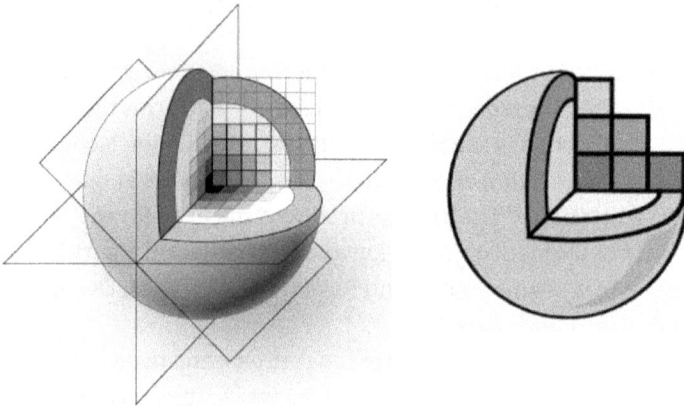

Slicer logo, 2018 and 2021

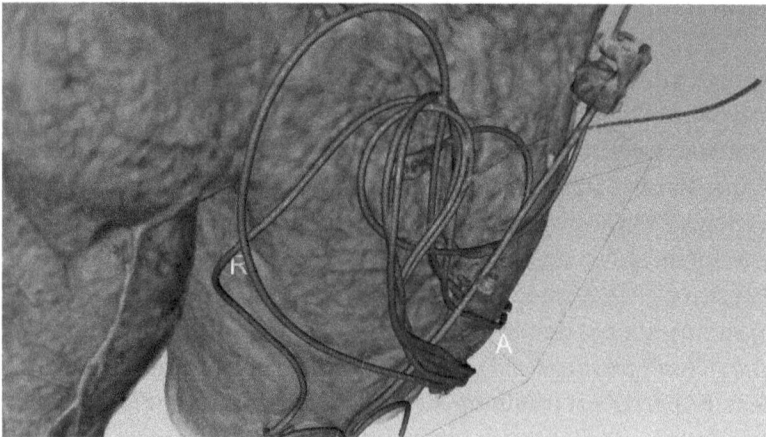

Re-rendered torso including medical equipment. Ray-tracing in Slicer v4.10.2 (screenshot)

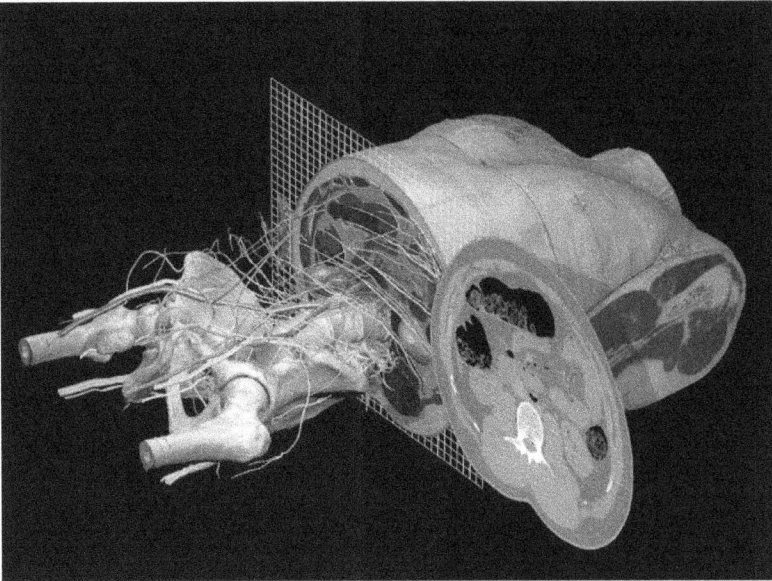

Torso of the Visible Human with a transverse section and a corresponding CT image, created with the VOXEL-MAN 3D atlas of anatomy and radiology; © VOXEL-MAN, University Medical Center Hamburg-Eppendorf, 2001, https://www.voxel-man.com

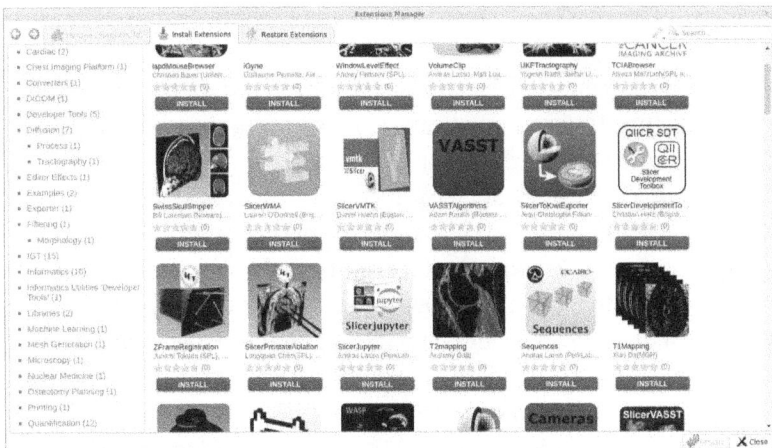

An abundance of extensions. Slicer v4.10.2 (screenshot)

3. Slicing

*Where we introduce the Modern Separation Toolkit, and
the aftermath of the cut.*

The act of separation is a key gesture of Modernity. The Modern Separation Toolkit (MST) contains persistent and culturally aligned modes of euro-andro-able-anthropocentric representation: taxonomy, anatomy, perspective, individual subjecthood, objectivity and many other material-semiotic moves of division. Separation is active on every level in order to isolate the part from the whole, the one from the other and to detach the object from the subject. Modern claims of truth work from the assumption that there is a necessary relation between separability, determinacy and sequentiality; between division, knowledge and representation.[27]

The disciplines of Art Theory, History of Science and the Philosophy of Perception exemplify with their individual means the particular gestures of separation in which the complexities of a particular world are haunted and caught by Modern modes to understand, name, transmit and eventually "apprehend" these worlds. If in tomography representing is a form of grasping or control, it is evident that we need to attend to the power relations that these cutting practices produce, so we don't allow them to be completely or definitively naturalized, culturally assumed as evident or given.

The specific mode of separation in contemporary biomedical imaging is the art of computational slicing. Our protagonist Slicer is obviously exposed to and exposing various cuts:

The subjectivity cut: Subjectivity can be understood as a prerequisite for representation, as it assures the presence of a subject responsible for a particular understanding of the world. But with the emergence of Modern subjecthood, of physical and legal persona freed from their environmental attachments and charged with free will and the capacity of judgment, additional representational norms impose themselves, somehow occupying an in-between space of singular and normative subjectivity.[28] In Slicer, the *subjectivity* cut is activated by the default choice of volumetric rendering, a two-point perspective where lines of sight come together in a single point, that of the individual viewer. These so-called bodies are reduced to their individual matter constellations, separated from the machinery around them, movable but divorced from their specific rhythms, without attachments or complications and most important of all, with minimal agency. Being and becoming is reduced to the incontestable promise of wholeness-at-the-end-of-the-scanner's-tunnel.

The regional cut refers to the techno-scientific phenomena of defining a Region of Interest (ROI), a location of special attention, even if it is as vast as a globe or an atlas. The regional cut supports a focus and a training of the gaze that as a result can habituate itself on a certain area, but only at the expense of not looking at another area.[29] In Slicer, the technical definition and subsequent isolation of what is called Region Of Interest operates as a computational upgrading of the decisions behind nineteenth century atlases of anatomy. This interface's operation presents the target as a cut. It results in a visual slicing of the virtual volume, which then exposes its invisible insides by its straight incisions.

The demarcation cut relates to the way that the practice of segmentation is present in both historical and contemporary biomedical imaging. Segmentation produces absolute divisions between image areas, organs, shades of gray and bones that obediently follow the anatomical canon. It all works together to give the renderings a sense of mathematical precision and medical evidence. In a nutshell, the process allows us to engineer a non-ambiguous spatial lay-out where each tissue or anatomical structure is identified by a label and a unique color code, all based on a black and white blur. The *demarcation cut* subsequently cascades into **The taxonomic cut** by means of the hierarchical anatomical model that Slicer shares with motion-tracking software.[30]

The invasive-non-invasive cut emerged when the tomographic paradigm imposed itself over other regimes of "seeing" in the field of biomedical imaging. This crossing concept connects the search for least invasivelessness in innovative surgery, with the thread of making invisible insides visible within biomedical informatics' research and practice. Slicer contributes to a dense constellation of techniques and technologies that are developed to cut bodies visually, but not in the flesh.

The last cut in this list is what we learned with Karen Barad to call **the agential cut**. They unfold a fundamental notion, that of intra-action, to give account of the constitutive onto-epistemes in apparatuses of observation. And this agential cut is fundamental for a trans*feminist approach to techno-sciences as response-ability.[31] This cut aims for a fundamental form of response-ability that is always already entangled in the production of knowledge and its apparatuses. In Slicer, we see the agential cut operating for example in the way the Open Source condition invites and expresses a mutual responsibility of users, devices, developers, algorithms, practitioners, researchers, datasets, founders, embodiments, and other involved agents.

These six cuts identify a number of agencies and their very particular distribution. Their power relations are based on aesthetic, economic and scientific paradigms which together define the tension between what is probable in the gesture of slicing, and what might be possible.

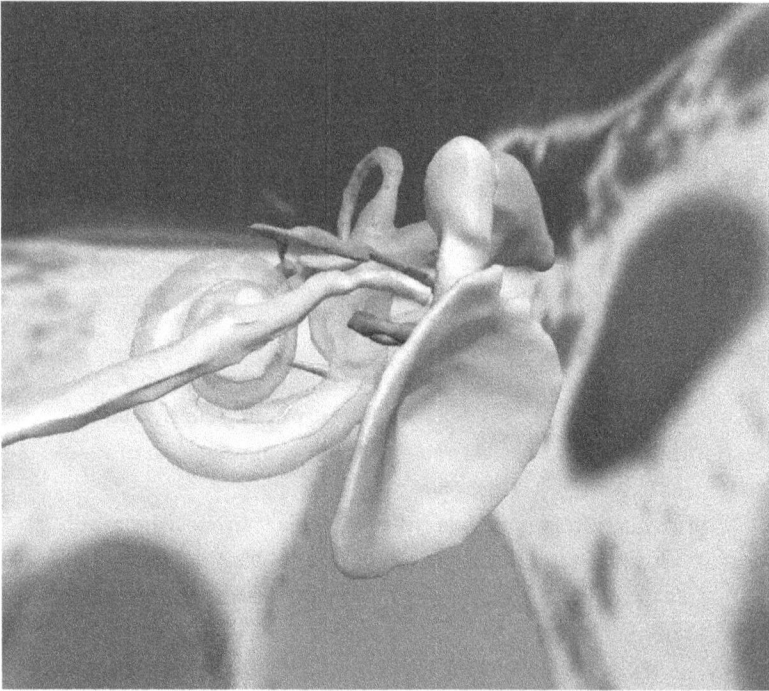

The demarcation cut: The SPL Inner Ear Atlas is based on CT scans visualized with Slicer. Open Anatomy Project. 2018, https://www.openanatomy.org

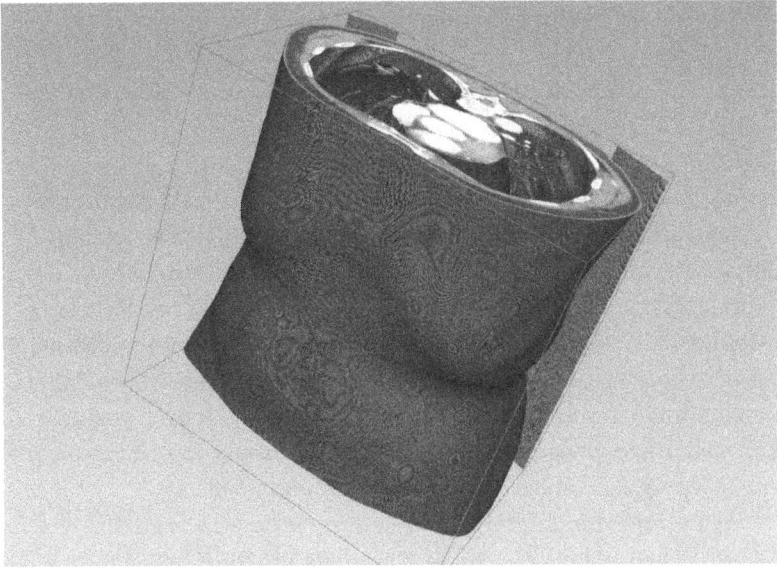

The regional cut: Defining a region of interest enacting a straight cut. Slicer v4.10.2 (screenshot)

Susan Potter's tissue was ground away, one hair-thin section — 63 microns — at a time.

The invasive-non-invasive cut: In 2015, Susan Potter donated her not-so normal body and her medical history to the Virtual Human Project. "How a Woman's Donated Body Became a Digital Cadaver," National Geographic YouTube channel, posted December 14, 2017, https://www.youtube.com/watc h?v=w-hhQNXQawU

4. Feature requests

Where the paradigmatic entanglement is ready to redistribute agencies.

In previous sections we moved from *slice* to *Slicer*, and then into *slicing*, encountering multiple entangled trans∗feminist urgencies on the way. We discussed the effects of the invention of the slice and the naturalization of its geometric and stratifying paradigms. We interrogated the agencies that altogether compose complex entanglements like our protagonist, Slicer. And in the last section, we listed six different cuts, understanding the act of division as a key Modern gesture that relates knowledge to (mostly visual) representation. Now it is time to apprehend Slicer's technicity by other means.[32]

With trans∗feminist techno-sciences we have learned that it is necessary to problematize Modern regimes and the impossibilities for life they produce. And that it is possible to do so with what we have at hand. Trans∗feminism challenges the ontology of humanity by questioning its separateness from social, economic, material, environmental, aesthetic and historical issues as well as from situated intersections such as race, gender, class, age, ability and species. They also invite us to test an ongoing *affirmative ethics*[33] in relation to the semiotic-material compositions of what we call "our worldings". It means to put ourselves at risk by reconsidering the very notion of "us", assuming the response-ability of being always already entangled with these techno-ecologies which we co-compose by just being-in-the-world.

Maybe Open Source platforms such as Slicer can be environments to render so-called bodies differently. Even if this software is being developed in the particularly tight hegemony of innovation-driven, biomedical research, its F/LOSS licensing conditions invites us to imagine an affirmative critique, in dialogue with the communities that develop the software. Or could the platform itself be rendered differently through disobedient takes on the body?

This text ends with a set of "feature requests" that challenge the slicedom of Slicer. It is an attempt at starting a kind of trans∗feminist prototyping for an open source software platform for biomedical informatics. To technically widen the tomographic imagination, we could maybe start by:

- Renaming the software platform to more accurately reflect the operations it performs. Some proposals: *Euclidean Anatomix, Forever dissecting, The Slicest, FlashFlesh, A-clinical Suite Pro, Tomographix Toolbox, Final Cut™, Kiss cut and the sensing knife...*[34]

- Introducing multiple and relational-perspectives. Computational rendering does not need a single vantage point, nor does it need to mimic the presence of human eyes. Next to the conventional two-way and orthogonal perspective, Slicer could bring multiple-axis and non-Euclidean perspective to the foreground.[35]

- De-centering the ocularcentrism of the renderings and re-orient representations. It is not (necessarily) about replacing vision with touch, vibrational, thermic and aural renderings although they might be less or otherwise burdened by Modern issues. We are first of all wondering about first of all collective modes of sensing and/or observation, to include multiplied modes of gathering and of processing impressions, of involving otherwise enabling renderings of data.

- Breaking the mirage of the interface as a mirror or window on a natural outcome. There must be ways to insist that representation is never complete: in volumetric renderings, nothingness and thereness are happening at the same time. Donna Haraway: "Partiality and not universality is the condition of being heard to make rational knowledge claims."[36]

- De-individualizing the imagery of the oneness of humanness. The platform does not need to technically collapse multiple slices into a discrete, single volumetric object that appear out of nowhere. Katherine Hayles says "only if one thinks of the subject as an autonomous self, independent of the environment, is one likely to experience the panic of Norbert Wiener's Cybernetics and Bernard Wolfe's Limbo [...] when the human is seen as part of a distributed system... it is not a question of leaving the body behind but rather of extending embodied awareness in highly specific, local and material ways that would be impossible without electronic prosthesis".[37]

- Problematzing the processual temporality of the volumetric images; can we make sure that we do not forget that these volumes as being constructed from takes at different moments, glued into a single object?

- Implementing *Agential Regions of Interest*. This is aimed at eventually liberating and freeing the slice from the Modern project. What would an a-Modern slice be, how would it behave? How to un-capture the slice from its Modern ghosts?

- Last but not least, we propose to dedicate some of funds to the initiation of a non-dependent program that would allow users, experts and other participants in Slicer to study the Computer Vision (sic) techniques that are implemented in this software. The program should not follow the limited spectrum of probable visions of a white-washed medical research imagination.

The possible is not about a fantastical widening of the imagination, but it is a technical condition that is already happening. This is a fundamental political twist in cultural analysis and critique of what imagination is: it is actually a technical thing. Imagination depends on the devices we collectively use, or that allow our lives to be used by. The devices we collectively use depend on that imagination. This dependency has always been and will always be *mutual*. When we assume this condition, then what would response-able imagery entail?

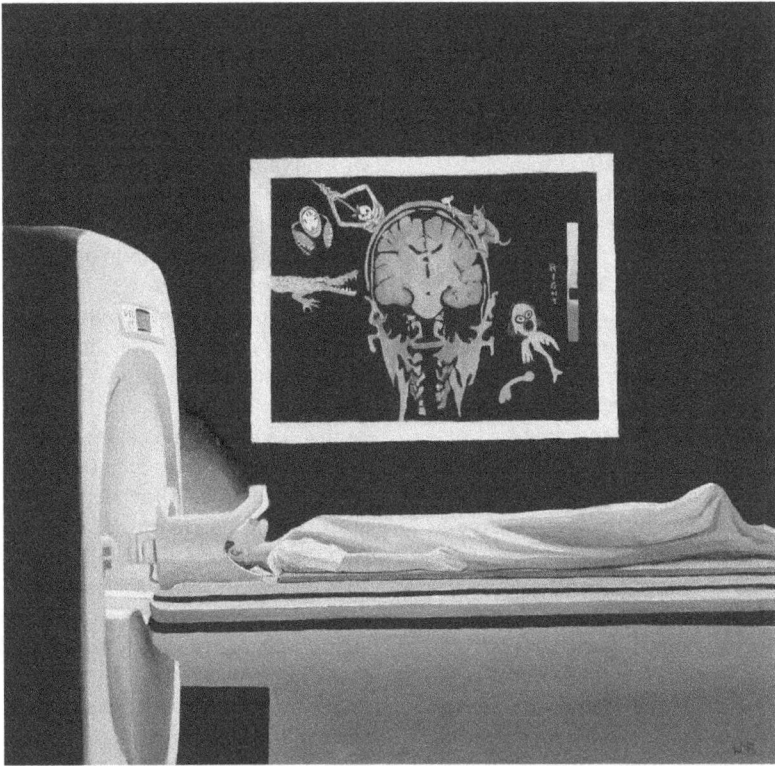

Lynn Randolph, "Immeasurable Result", oil on masonite, 1994. Included in Donna J. Haraway, *Modest_Witness@Second_Millennium. FemaleMan©_Meets_OncoMouse™. Feminism and Technoscience*, Originally published in 1997.

Notes

1. ↑ We apply the formula *trans∗feminist* in order to convoke all necessary intersectional and intrasectional aspects around that star (∗).

2. ↑ Computer Tomography (CT) uses multiple x-ray-exposures; Positron-Emission Tomography (PET) reads from radioactive tracers that a subject has swallowed or was injected with and Magnetic Resonance Imaging (MRI) uses strong magnets and then measures the difference in speed between activation and dis-activation of atoms.

3. ↑ Lorraine Daston, and Peter Galison, "The image of objectivity," *Representations*, No. 40, Special Issue: *Seeing Science* (Autumn, 1992): 106.

4. ↑ In 1917, Austrian mathematician Johann Radon introduced the Radon transform, a formula that fifty years later Sir Godfrey Hounsfield would combine with a quantitative scale for radiodensity, the Hounsfield unit (HU), to reverse-calculate images from density projection data in the CT-scanner that he invented.

5. ↑ In 2017 circa 13.000 CT-scanners in European hospitals performed 80 million scans per year. See: "Healthcare resource statistics – technical resources and medical technology Statistics Explained," Eurostat, 2019, https://ec.europa.eu/eurostat/statistics-explained/pdfscache/37388.pdf.

6. ↑ See: "Item 074: The Continuum," *The Possible Bodies Inventory*, 2017.

7. ↑ CT-scanners are not non-invasive at all since they use x-rays that carry a risk of developmental problems and cancer. This triggered for example *Image Gently*, a campaign to be more careful with radiation especially when used on children. https://www.imagegently.org.

8. ↑ Sara Ahmed, *Queer Phenomenology, Orientations, Objects, Others* (Durham: Duke University Press, 2006), 70.

9. ↑ Euclidian geometry relies among others on the parallel postulate: "if a straight line falling on two straight lines make the interior angles on the same side less than two right angles, the two straight lines, if produced indefinitely, meet on that side on which the angles are less than two right angles." "Euclidean Geometry," Wikipedia, accessed October 20, 2021, https://en.wikipedia.org/wiki/Euclidean_geometry.

10. ↑ "Through the dissection and analysis of the body's organisation, anatomy works to suspend any distinction between surface and depth, interior and exterior, endosoma and exosoma. It ideally makes all organs equally available to instrumental address and calibration, forms of engineering and assemblage with other machine complexes." Catherine Waldby, *The Visible Human Project: Informatic Bodies and Posthuman Medicine* (Milton Park: Routledge, 2000), 51.

11. ↑ "The woman lies comfortably relaxed; the artist sits upright, rigidly constrained by his fixed position. The woman knows that she is seen; the artist is blinded by his viewing apparatus, deluded by his fantasy of objectivity. The draftsman's need to order visually and to distance himself from that which he sees suggests a futile attempt to protect himself from what he would (not) see. Yet the cloth draped between the woman's legs is not protection enough; neither the viewing device nor the screen can delineate or contain his desire. The perspective painter is transfixed in this moment, paralyzed, unable to capture the sight that encloses him. Enclosing us as well, Dürer's work draws our alarm." Barbara Freedman, *Staging the Gaze: Postmodernism, Psychoanalysis, and Shakespearean Comedy* (Ithica: Cornell University Press, 1991), 2.

12. ↑ W.E. Lorensen and Harvey Cline, "Marching cubes: A high resolution 3d surface construction algorithm," *ACM Computer Graphics* 21 (1987): 163–169.

13. ↑ See Karen Barad, "Getting Real: Technoscientific practices and the materialization of reality," in: *Meeting the Universe Halfway*, (Durham: Duke University Press, 2007), 189-222.

14. ↑ Aud Sissel Hoel, and Frank Lindseth, "Differential Interventions: Images as Operative Tools," in *Photomediations: A Reader*, eds. Kamila Kuc and Joanna Zylinska (Open Humanities Press, 2016), 177-183.

15. ↑ Slicer documentation, download and forum pages each describe its main purpose in slightly different ways: "an open source software platform for medical image

informatics, image processing, and three-dimensional visualization." https://www.slicer.org/wiki/Main_Page. "Slicer, or 3D Slicer, is a free, open source software package for visualization and image analysis." https://github.com/Slicer/Slicer. "3D Slicer ("Slicer") is an open source, extensible software platform for image visualization and analysis. Slicer has a large community of users in medical imaging and surgical navigation, and is also used in fields such as astronomy, paleontology, and 3D printing." https://discourse.slicer.org/t/slicer-4-8-summary-highlights-and-changelog/1292. "a software platform for the analysis (including registration and interactive segmentation) and visualization (including volume rendering) of medical images and for research in image guided therapy," https://slicer.readthedocs.io/en/latest/user_guide/getting_started.html. All URLs accessed July 1, 2021.

16. ↑ Waldby, *The Visible Human Project*, 34.

17. ↑ The Slicer publication database hosted by the Surgical Planning Laboratory currently contains 552 publications, http://www.spl.harvard.edu/publications/pages/display/?collection=11.

18. ↑ When launching Slicer, a pop-up appears: "This software is not intended for clinical use". In the main interface we also find "This software has been designed for research purposes only and has not been reviewed or approved by the Food and Drug Administration, or by any other agency." In addition, the software license stipulates in capital letters that "YOU ACKNOWLEDGE AND AGREE THAT CLINICAL APPLICATIONS ARE NEITHER RECOMMENDED NOR ADVISED,"

accessed October 1, 2021, https://github.com/Slicer/Slicer/blob/master/License.txt

19. ↑ Slicer positions itself as a prototyping environment in-between diagnostic practice and innovative marketable products, and "facilitates translation and evaluation of the new quantitative methods by allowing the biomedical researcher to focus on the implementation of the algorithm, and providing abstractions for the common tasks of data communication, visualization and user interface development." Andriy Fedorov et al., "3D Slicer as an image computing platform for the Quantitative Imaging Network," *Magnetic resonance imaging* vol. 30, no. 9 (2012): 1323-41.

20. ↑ Tina Kapur et al. "Increasing the impact of medical image computing using community-based open-access hackathons: The NA-MIC and 3D Slicer experience," *Medical Image Analysis* 33 (2016): 176-180.

21. ↑ "The term 'cut' is a bit of a misnomer, yet it is used to describe the process of grinding away the top surface of a specimen at regular intervals. The term 'slice,' also a misnomer, refers to the revealed surface of the specimen to be photographed; the process of grinding the surface away is entirely destructive to the specimen and leaves no usable or preservable 'slice' of the cadaver." "The Visible Human Project," Wikipedia, accessed October 20, 2021, https://en.wikipedia.org/wiki/Visible_Human_Project.

22. ↑ Naming is a strongly politicized representational technique. See also Paul B Preciado, *Testo Junkie: Sex, Drugs, and Biopolitics in the Pharmacopornographic Era* (New York: Feminist Press, 2013) for a

discussion of the theological-patriarchal regime of the biomedical field.

23. ↑ See Ursula K. Le Guin, "She unnames them," or "Item 059: Anarcha's Gland," *The Possible Bodies Invenotory*, for an account of the attempt by tech-feminist group Pechblenda to rename anatomy in an attempt to decolonize bodies.

24. ↑ "The Visible Human Project data sets are designed to serve as a common reference point for the study of human anatomy, as a set of common public domain data for testing medical imaging algorithms, and as a test bed and model for the construction of image libraries that can be accessed through networks." "Programs and services fiscal year 2000," National Institutes of Health, National Library of Medicine, 2000, https://www.nlm.nih.gov/ocpl/anrep orts/fy2000.pdf.

25. ↑ "Insight Segmentation and Registration Toolkit," accessed October 1, 2021, https://itk.org/Doxygen413/html/ind ex.html

26. ↑ "Technics can intensify and multiply force and forms of vitality by ordering it as an economy, a calculable and hierarchical system of value – exist in circulation and disctribution, can function in other economies." Waldby, *The Visible Human Project*, 33.

27. ↑ As Rosi Braidotti notes, "Modern science is the triumph of the scopic drive as a gesture of epistemological domination and control: to make visible the invisible, to visualise the secrets of nature. Biosciences achieve their aims by making the embodied subject visible and intelligible according to the principles of scientific representation. In turn this implies that the body can be split into a variety of organs, each of which can be analyzed and represented." Rosi Braidotti, *Nomadic Subjects: Embodiment and Sexual Difference in Contemporary Feminist Theory* (New York: Columbia University Press, 2011), 196.

28. ↑ Daston, and Galison, "The image of objectivity," 106.

29. ↑ "[W]hat was not new to nineteenth-century atlases was the dictum "truth to nature": there is no atlas in any field that does not pique itself on its accuracy, on its fidelity to fact. But in order to decide whether an atlas picture is an accurate rendering of nature, the atlas maker must first decide what nature is. All atlas makers must solve the problem of choice: which objects should be presented as the standard phenomena of the discipline, and from which viewpoint? In the late nineteenth century, these choices triggered a crisis of anxiety and denial, for they seemed invitations to subjectivity." Daston, and Galison, "The image of objectivity," 86.

30. ↑ The model for anatomical data in Slicer resembles the crude cascading hierarchies used in basic motion tracking software.

31. ↑ "We are responsible for the world within which we live not because it is an arbitrary construction of our choosing, but because it is sedimented out of particular practices that we have a role in shaping. and 'The crucial point is that the apparatus enacts an agential cut – a resolution of the ontological indeterminacy – within the phenomenon, and agential separability – the agentially enacted material condition of exteriority-within-phenomena – provides the condition for the possibility of objectivity. This *agential cut* also

instrument (effect) by the measured object (cause), where 'local' means within the phenomenon." Karen Barad, "Getting Real," 390 & 175.

32. ↑ Hoel, "Differential Interventions: Images as Operative Tools," 177-183.

33. ↑ Rosi Braidotti, "Affirmative Ethics, Posthuman Subjectivity, and Intimate Scholarship: a Conversation with Rosi Braidotti", in "Decentering the Researcher in Intimate Scholarship", *Advances in Research on Teaching*, no. 31, Emerald Publishing Limited, 2018, 179-188.

34. ↑ "Only our starfish can save us, by regrowing whatever grooms like me cut out of them. Grandma Chan Ling invented the kiss cut, the repair job — what do you say? The fix, the patch. The first starfish gave her liver, her kidneys, and, at last, her red-hot heart to the first doubler. And so it was, in the beginning." Larissa Lai, *The Tiger Flu*, 2019, or see: Item 116: *Kiss cut and the sensing knife,* The Possible Bodies inventory, 2020.

35. ↑ Slicer does offer a second perspective rendering, namely "orthographic perspective" (straight-extreme).

36. ↑ Donna Haraway, "Situated Knowledges: The Science Question in Feminism and the Privilege of Partial Perspective," *Feminist Studies* 14, no.3 (1998): 589.

37. ↑ N. Katherine Hayles, "What does it mean to be posthuman?," in *The New Media and Cybercultures Anthology*, ed. Pramod K. Nayar (Hoboken: Wiley-Blackwell, 1988), 25.

Parametric Unknowns: Hypercomputation between the probable and the possible

Panoramic Unknowns

Nicolas Malevé

1. The Caltech's lab seen by a Kodak DC280

I am looking at a folder of 450 pictures. This folder named faces1999 is what computer vision scientists call a dataset, a collection of images to test and train their algorithms. The pictures have all been shot using the same device, a Kodak DC280, a digital camera aimed at the "keen amateur digital photographer".[1] If the Kodak DC280 promised a greater integration of the camera within the digital photographic workflow, it was not entirely seamless and required the collaboration of the photographer at various stages. The camera was shipped with a 20 MB memory card. The folder size is 74.8 MB, nearly four times the card's storage capacity. The photographs have been taken during various sessions between November 1999 and January 2000 and transferred to a computer to empty the card several times. Additionally, if the writing on the card was automatic, it was not entirely transparent. As product reviewer Phil Askey noted, "Operation is quick, although you're aware that the camera takes quite a while to write out to the CF card (the activity LED indicates when the camera is writing to the card)."[2]

Moving from one storage volume (the CF card) to another (the researcher's hard drive), files acquire a new name. A look at the file names in the dataset reveals that the dataset is not a mere dump of the successive shooting sessions. By default, the camera follows a generic naming procedure: the photos' names are composed of a prefix "dcp_" followed by a five digit identifier padded with zeroes (ie. dcp_0001.jpg, dcp_0002.jpg, etc.). The photographer however took the pain of renaming all the pictures following his own convention, he used the prefix "image_" and kept the sequential numbering format (ie. image_0001.jpg, image_002.jpg, etc). The photo's metadata shows that there are gaps between various series of shots and that the folder's ordering doesn't correspond to the image's capture date. It is therefore difficult to say how far the photographer went into the reordering of his images. The ordering of the folder has erased the initial ordering of the device, and some images may have been discarded.

The decision to alter the ordering of the photos becomes clearer when observing the preview of the folder on my computer. My file manager displays the photos as a grid, offering me a near comprehensive view of the set. What stands out from the ensemble is the recurrence of the centered frontal face. The photos are ordered by their content, the people they represent. There is a clear articulation between figure and background, a distribution of what the software will need to detect and what it will have to learn to ignore.[3] To enforce this division, the creator of the data set has annotated the photographs: in a file attached to the photographs, he provides the coordinates of the faces represented in the photos. This foreground/background division pivoting on the subject's face relates to what my interlocutors, Femke and Jara whose commentaries and writings are woven within this text, are calling a *volumetric regime.* This expression in our conversations functions as a sensitizing device to the various operations of naturalized volumetric and spatial techniques. I am refraining from defininge it now and will provisionally use the expression to signal, in this situation, the preponderance of an organising pattern (face versus non-face) implying a planar hierarchy. Simultaneously, this first look at the file manager display generates an opposite sensation: the intuition that other forms of continuity are at play in the data set. This

106

complicates what data is supposed to be and the web of relations it is inserted in.

2. Stitching with Hugin

The starting point of this text is to explore this intuition: is there a form of spatial trajectory in the data set and how to attend to it? I have already observed that there was a spatial trajectory inherent in the translation of the files from storage volume to another. This volumetric operation had its own temporality (ie. unloading the camera to take more photos), it brought in its own nomenclature (renaming of the files and its re-ordering). The spatial trajectory I am following here is of another nature. It happens when the files are viewed as photographs, and not as merely as arrays of pixels. It is a trajectory that does not follow the salient features the data set is supposed to register, the frontal face. Instead of apprehending the data set as a collection of faces, I set out to follow the trajectory of the photographer through the lab's maze. *Faces1999* is not spatially unified, it is the intertwining of several spaces: offices, corridors, patio, kitchen ... more importantly, it conveys a sense of provisional continuity and passage. How to know more about this intuition? How to find a process that sets my thoughts in motion? As a beginning, I am attempting to perform what we call a *probe* at the Institute for Computational Vandalism: pushing a software slightly outside of its boundaries to gain knowledge about the objects it takes for granted.[4] In an attempt to apprehend the spatial continuum, I introduce the data set's photographs in an image panorama software called Hugin. I know in advance that using these photos as an input for Hugin will push the boundaries of the software's requirements. The ideal scenario for software such as Hugin is a collection of photographs taken sequentially and its task is to minimise the distortions produced by changes in the point of view. For Hugin, different photos can be aligned and re-projected on the same plane. The software won't be able to compensate for the incompleteness of the spatial representation, but I am interested to see what it does with the continuities and contiguities, even as partial as they are. I am interested to follow its process and to see where it guides my eyes.

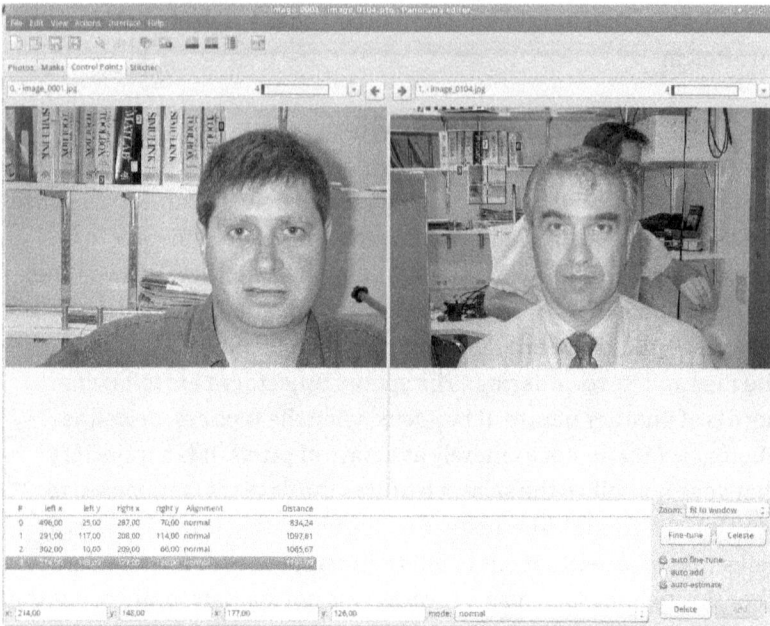

Two images in Hugin

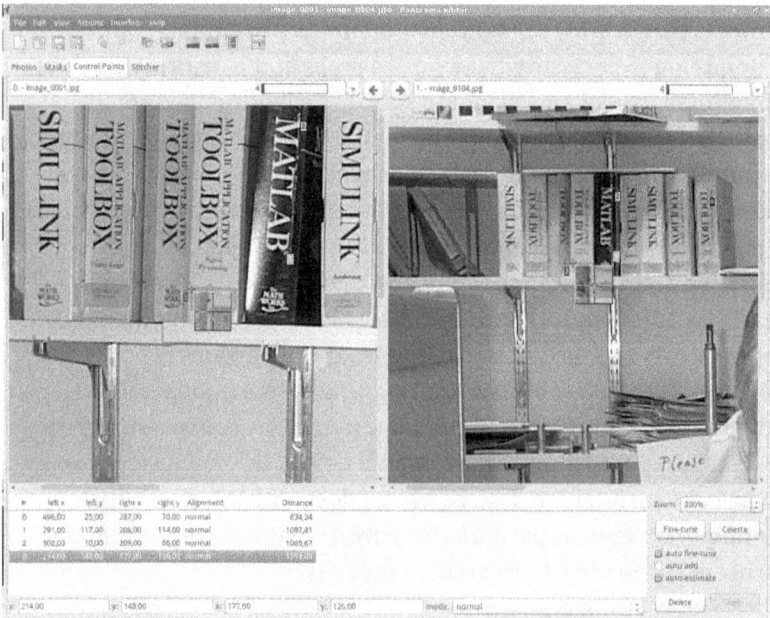

Selecting points of interest manually

Hugin can function autonomously and look for the points of interest in the photographs that will allow it to stitch the different views together. It can also let the user select these points of view. For this investigation, the probe is made by a manual selection of the points in the backgrounds of the photos. To select these points, I am forced to look for the visual clues connecting the photos. Little by little, I reconstruct two bookshelves forming a corner. Then, elements in the pictures become eventful. Using posters on the wall, I discover a door opening onto an office with a window with a view onto a patio. Comparing the orientation of the posters, I realize I am looking at different pictures of the same door. I can see a hand, probably of a person sitting in front of a computer. As someone shuts the door, the hand disappears again. One day later, the seat is empty, books have been rearranged on the shelves, stacks of papers have appeared on a desk. In two months, the backgrounds slowly moves, evolves. On the other side of the shelves, there is a big white cupboard with an opening through which one can see a slide projector. Following that direction, is a corridor. The corridor wall is covered with posters announcing computer vision conferences and competitions for students. There is also a selection of photographs representing a pool party that helps me "articulate" several photographs together. Six pictures show men in a pool. Next to these, a large photo of a man laying down on the grass in natural light, and is vaguely reminiscent of an impressionist painting. Workers partying outside of the workplace pictured on the workplace's walls.

Hugin trying to resolve different view points

Background prominence

At regular intervals, I press a button labeled "stitch" in the panorama software and Hugin generates for me a composite image. Hugin does not merely overlay the photos. It attempts to correct the perspectival distortions, smooth out the lighting contrasts, resolve exposure conflicts and blend the overlapping photos. When images are added to the panorama, the frontal faces are gradually faded and the background becomes salient. As a result, the background is transformed. Individual objects loose their legibility, books titles fade. What becomes apparent is the rhythm, the separations and the separators, the partition of space. The material support for classification takes over its content: library labels, colors of covers and book edges become prominent.

Finding a poster in a photo, then seeing it in another photo, this time next to a door knob, then in yet another that is half masked by another poster makes me go through the photos back and forth many times. After a while, my awareness of the limits of the corpus of photos grows. It grows enough to have an incipient sensation of a place out of the fragmentary perceptions. And concomitantly, a sense of the missing pictures, missing from a whole that is nearly tangible. With a sense that their absence can be perhaps compensated. Little by little, a traversal becomes possible for me. Here, however Hugin and I part ways. Hugin gives up on the overwhelming task of resolving all these views into a coherent perspective. Its attempt to recover the contradictory perspectives ends up in a flamboyant spiraling outburst. Whilst Hugin attempts to close the space upon a spherical projection, the tedious work of finding connecting points in the photos gave me another sensation of the space, passage by passage, abandoning the idea of a point of view that would offer an overarching perspective. Like how a blind person touching the contiguous surface can find their way through the maze, I can intuit continuities, contiguities, and spatial proximities that open a volume onto one another. The dataset opens up a world with depth. There is a body circulating in that space, the photos are the product of this circulation.

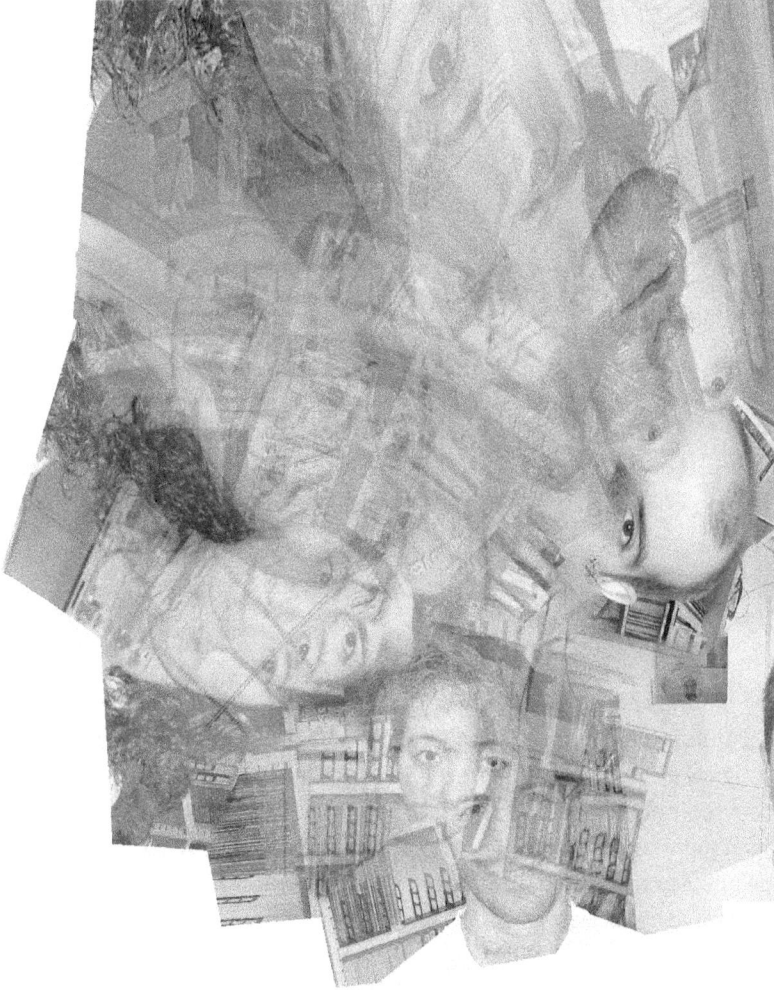

Spiralling outburst

Spiralling outburst

3. Accidental ethnography

As I mentioned at the beginning of this text, Griffin this folder of photographs is what computer vision engineers call a dataset: a collection
of digital photographs that developers use as material to test and
train their algorithms on. Using the same dataset allows different
developers to compare their work. The notice that comes along with
the photographs gives a bit more information about the purpose of
this image collection. The notice, a document named README, states:

> *Frontal face dataset. Collected by Markus Weber at*
> *California Institute of Technology.*
>
> *450 face images. 896 x 592 pixels. Jpeg format.*
> *27 or so unique people under with different*
> *lighting/expressions/backgrounds.*

ImageData.mat is a Matlab file containing the variable
SubDir_Data which is an 8 x 450 matrix.
Each column of this matrix hold the coordinates of the
bike within the image, in the form:

[x_bot_left y_bot_left x_top_left y_top_left ... x_top_right
y_top_right x_bot_right y_bot_right]

R. Fergus 15/02/03

As announced in the first line, *Faces1999* contains pictures of people
photographed frontally. The collection contains mainly close-ups of
faces. To a lesser degree, it contains photographs of people in medium
shots. And to an even less degree, it contains three painted silhouettes
of famous actors, like of Buster Keaton. But my trajectory with Hugin,
my apprehension of stitches and passages leads me elsewhere than
the faces. I am learning to move across the dataset. This movement is
not made of a series of discrete steps, each positioning me in front of
a face (frontal faces) but a transversal displacement. It teaches me to
observe textures and separators, grids, shelves, doors, it brings me
into an accidental ethnography of the lab surfaces.

Most of the portraits are taken in the same office environment. In
the background, I can see shelves stacked with programming books,
walls adorned with a selection of holiday pictures, an office kitchen,
several white boards covered with mathematical notations, news
boards with invitations to conferences, presentations, parties, and
several files extracted from a policy document, a first aid kit next to a
box of Nescafé, a slide projector locked in a cupboard.

Looking at the books on display on the different shelves, I play
with the idea of reconstructing the lab's software ecosystem. Software
for mathematics and statistics: thick volumes of Matlab and Matlab
related manuals (like Simulink), general topics like vector calculus,
applied functional analysis, signal processing, digital systems engineering, systems programming, concurrent programming, specific
algorithms (active contours, face and gesture recognition, the EM
algorithm and extensions) or generic ones (a volume on sorting and

searching, cognition and neural networks), low level programming languages Turbo C/C++, Visual C++ and Numerical Recipes in C. Heavily implanted in math more than in language. The software ecosystem also includes resources about data visualisation and computer graphics more generally (the display of quantitative information, Claris Draw, Draw 8, OpenGL) as well as office related programmes (MS Office, Microsoft NT). There are various degrees of abstraction on display. Theory and software manuals, journals, introductions to languages and specialized literature on a topic. Book titles ending with the word theory or ending with the word "programming", "elementary" or "advanced". Design versus recipe. There is a mix of theoretical and applied research. The shelves contain more than software documentation: an electronic components catalogue and a book by John Le Carré are sitting side by side. It Ironically reminds us that science is not made with science only, neither software by code exclusively.

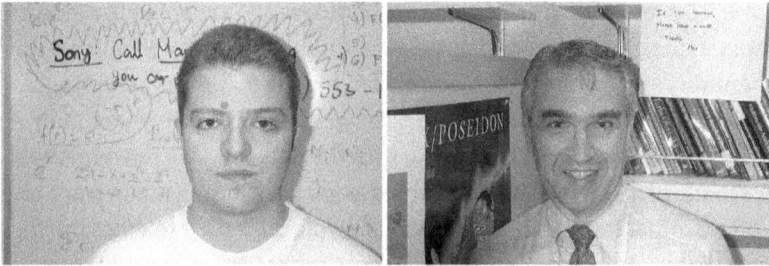

Inscriptions. The Computational Vision Group, *Faces1999 (Front)*, 1999. Accessed February 1, 2022, http://www.vision.caltech.edu/archive.html.

Books are stacked. Each book claiming its domain. Each shelf adding a new segment to the wall. Continuing my discovery of spatial continuities, I turn my attention to surfaces with a more conjunctive quality. There is a sense of conversation happening in the backgrounds. The backgrounds are densely covered with inscriptions of different sorts. They are also overlaid by commentaries underlying the mixed nature of research activity. Work regulation documents (a summary of the Employee Polygraph Protection Act), staff emails, address directories, a map of the building, invitations to conferences and parties, job ads, administrative announcements, a calendar page for October 1999, all suggest that more than code and mathematics are happening in this environment. These surfaces are calling their readers out: bureaucratic injunctions, interpellations, invitations using the language of advertising. On a door, a sign reads "Please do not disturb". A note signed Jean-Yves insists "Please do NOT put your fingers on the screen. Thanks." There are networks of colleagues in the lab and beyond. These signs are testament to activities they try to regulate: people open doors uninvited and show each other things on screens leaving finger traces. But the sense of intertwining of the ongoing social activity and the work of knowledge production is nowhere more present than in the picture of a whiteboard where complex mathematical equations cohabit with a note partially masked by a frontal face: Sony Call Ma... your car is... 553-1. The same surface of inscription is used for both sketching the outline of an idea and internal communication.

Approaching the dataset this way offers an alternative reading of the manners in which this lab of computer vision represents itself and to others what its work consists of. The emic narrative doesn't offer a mere definition of the members activity. It comes with its own continuities. One such continuity is the dataset's temporal inscription into a narrative of technical progress that results in a comparison with the current development of technology. I realize the difficulty to resist it. How much I myself mentally comparei the Kodak camera to the devices I am using. I take most of m Griffin y photos with a phone. My phone's memory card is 10 gigabytes whereas Kodak proudly advertised a 20MB card for its DC280 model. The dataset's size pales in comparison to current standards (a state-of-the-art dataset as UMD-Faces includes 367,000 face annotations[5] and VGGFace2 provides 3,3 million face images downloaded from Google Image Search).[6] The question of progress here is problematic in that it tells a story of continuity that is recurrent in books, manuals and blogs related to AI and machine learning. This story can be sketched as: "Back in the days, hardware was limited, data was limited, then came the data explosion and now we can make neural networks properly."[7] Whilst this narrative is not inherently baseless, it makes it difficult to attend to the specificity of what this dataset is and how it relates to larger networks of operation. And what can be learned from it. In a narrative of progress, it is defined by what it is no longer (it is not defined by the scarcity of the digital photograph anymore) and by what it will become (*faces1999* is like a contemporary dataset, but smaller). The dataset is understood through a simple narrative of volumetric evolution where the exponential increase of storage volumes rhymes with technological improvement. Then it is easy to be caught in a discourse that treats the form of its photographic elaboration as an in-between. Already digital but not yet networked, post-analogue but pre-Flickr.

4. Photography and its regular objects

So how to attend to its photographic elaboration? What are the devices and the organisation of labour necessary to produce such a thing as *Faces1999*? The photographic practice of the Caltech engineers matters more than it may seem. Photography in a dataset such as this is a leveler. It is the device through whichdisparate fragments that make up the visual world can be compared. Photography is used as a tool for representation and as a tool to regularise data objects. The regularization of scientific objects opens the door to the representation and naturalization of cultural choices. It is representationally active. It involves the encoding of gender binaries, racial sorting, spatial delineation (what happens indoors and outdoors in the dataset). Who takes the photo, who is the subject? Who is included and who is excluded? The photographer is a member of the community. In some way, *he*[8] is the measure of the dataset. It is a dataset at his scale. To move through the dataset is to move through his spatial scale, his surroundings. Where he can easily move and recruit people, he has bonds with the "subjects". He can ask them "come with me", "please smile" to gather facial expressions. Following the photographer, we move from the lab to the family circle. About fifty photos interspersed with the lab photos represent relatives of the researchers in their interior spaces. While it is difficult to say for certain how close they are, they depict women and children in a house interior. It is his world, ready to offer itself to his camera.

Further, to use Karen Barad's vocabulary, the regularization performs an agential cut: it enacts entities with agency and by doing so, it enforces a division of labor.[9] My characterization of the photographer and *his* subject has until now has remained narrow. The subjects do not simply respond to the photographer, but they respond to an assemblage comprising at a minimum the photographer, camera, familiar space, lighting condition, and storage volumes. To take a photo means more a than a transaction between a person seeing and a person seen. Proximity here does not translate smoothly to intimacy. In some sense, to regularize his objects, as a photographer, the dataset maker must be like everyone else. The photograph must be at some level interchangeable with those of the "regular photographer". The procedure to acquire the photographs of *faces1999* is not defined.

Yet regularization and normalization are at work. The regulative and normalizing functions of the digital camera, its ability to adapt, its distribution of competences, its segmentation of space are operating. But also its conventions, its acceptability. The photographic device here works as a soft ruler that adjusts to the fluctuating contours of the objects it measures.

Its objects are not simply the faces of the people in front of the photographer. The dataset maker's priority is not to ensure indexicality. He is less seeking to represent the faces as if they were things "out there" in the world than trying to model a form of mediation. The approach of the *faces1999* researchers is not one of direct mediation where the camera simply is considered a transparent window to the world. If it were the case, the researchers would have removed all the "artefactual" photographs wherein the mediation of the camera is explicit: where the camera blurs or outright cancels the representation of the frontal face. What it models instead is an average photographic output. It does not model the frontal face, it models the frontal face as mediated by the practice of amateur photography. In this sense, it bears little relation with the tradition of scientific photography that seeks to transparently address its object. To capture the frontal face as mediated by vernacular photography, the computer scientist doesn't need to work hard to remove the artifactuality of its representation. He needs to work as little as possible, he needs to let himself guided by a practice external to his field, to let vernacular photography infiltrate his discipline.

Backlighting, overexposed. The Computational Vision Group, *Faces1999 (Front)*, 1999. Accessed February 1, 2022, http://www.vision.caltech.edu/archiv e.html.

The dataset maker internalizes a common photographic practice. For this, he must be a particular kind of functionary of the camera as Flusser would have it.[10] He needs to produce a certain level of entropy in the program of the camera. The camera's presets are determined to produce predictable photographs. The use of the flash, the speed, the aperture are controlled by the camera to keep the result within certain aesthetics norms. The regularization therefore implies a certain dance with the kind of behavior the photographer is expected to adopt. If the dataset maker doesn't interfere with the regulatory functions of the camera, the device may well regularize too much of the dataset and therefore move away from the variations that one can find in amateur photo albums. The dataset maker must therefore trick the camera to make enough "bad" photos as would normally happen over the course of a long period of shooting. The flash must not fire at times even when the visibility in the foreground is low. This requires the circumvention of the default camera behavior in order to provoke an accident bound to happen over time. A certain amount of photos must be taken with the subject off-centre. Faces must be occasionally out of focus. And when an accident happens by chance, it is kept in the dataset. However, these accidents cannot exceed a certain threshold: images need to remain generic. The dataset maker explores the thin range of variation in the camera's default mode that corresponds to the mainstream use of the device. The researcher does not systematically explore all the parameters. They introduce a certain wavering in its regularities. A measured amount of bumps and lumps. A homeopathic dose of accidents. At each moment, there is a perspective, a trajectory that inflects the way the image is taken. It is never only a representation, it always anticipates variations and redundancies, it always anticipates its ultimate stabilization as data. The identification of exceptions and the inclusion of accidents is part of the elaboration of the rules. The dataset maker cannot afford to forget that the software does not need to learn to detect faces in the abstract. It needs to learn to detect faces as they are made visible within a specific practice of photography and internalized to some degree by the camera.

5. Volumetric regimes

Everything I have written until now has been the result of several hours of looking at the *faces1999* images. I have done it through various means. In a photo gallery, through an Exif reader program, via custom code, and through the panorama software Hugin. However, nowhere in the README or the website where the dataset can be downloaded, can an explicit invitation to look at the photos be found.

The README refers to one particular use. The areas of interest compiled in the Matlab file makes clear that the privileged access to the dataset is through programs that treat the images as matrices of numbers and process them as such. It doesn't mean a dataset such as *faces1999* cannot be treated as an object to be investigated visually. Twenty-seven faces is an amount that one person can process without too much trouble. One can easily differentiate them and remember most of them. For the photographer and the person who annotated the dataset, traced the bounding boxes around the faces, the sense of familiarity was even stronger. They were workmates or even family. The dataset maker could be present at all stages in the creation of the dataset: he would select the people, the backgrounds, press the shutter, assemble and rename the pictures, trace the bounding boxes, write the readme, compress the files and upload them on the website. Even if Hugin could not satisfactorily resolve the juxtaposition of points of view, its failure still hinted at a potential panorama ensuring, a continuity through the various takes. There was at least a possibility of an overview, of grasping a totality.

This takes me back to the question of *faces1999*'s place in a narrative of technological progress. In such narrative, it plays a minor role and should be forgotten. It is not a standard reference of the field and its size pales in comparison to current standards. However, my aim with this text is to insist that datasets in computer vision should not be treated as mere collections of data points or representations that can be simply compared quantitatively. They articulate different dimensions and distances. If the photos cut the lab into pieces, to assemble *faces1999* implies a potential stitching of these fragments. This created various virtual pathways through the collection that mobilized conjunctive surfaces, walls covered of instructions and recursive openings (door opening on an office with a window opening on a patio). There were passageways connecting the lab to the home and back. There was cohesion if not coherence. At the invitation of Jara and Femke, taking the idea of a volumetric regime as a device to think together the sequencing of points of views, the naturalization of the opposition between face and background, the segmentations, but also the stitches, the passageways, the conjunctive surfaces, the storage volumes (of the brand new digital camera and the compressed archive) through which the dataset is distributed, I have words to apprehend better the singularity of *faces1999*. *Faces1999* is not a small version of a contemporary dataset. A quantitative change reaches out into other dimensions, another space, another coherence, another division of labour and another photographic practice. Another volumetric regime.

Acknowledging its singularity does not mean to turn *faces1999* into a nostalgic icon. It matters because recognizing its volumetric regime changes the questions that can be asked of current datasets. Instead of asking how large they are, how much they have evolved, I may be asking to which volumetric regime they belong (and they help enact in return). Which means a flurry of new questions need to be raised: what are the dataset's passageways? How do they split and stitch? What are its conjunctive surfaces? What is the division of labour that subtends it? How is the photographic apparatus involved in the regularization of their objects? And what counts as photographic apparatus in this operation?

Asking these questions to datasets such as MegaFace, Labelled Faces in the wild or Google facial expression comparison would immediately signal a different volumetric regime that cannot be reduced to a quantitative increase. The computer scientist once amateur photographer becomes photo-curator (the photos are sourced from search engines rather than produced by the dataset maker). The conjunctive surfaces that connect administrative guidelines and mathematical formulas would not be represented in the photos backgrounds but built into the contract and transactions of the platform of annotation that recruits the thousands of workers necessary to label the images (instead of the lone packager of *faces1999*). Their passageways should not be sought in the depicted spaces in which the faces appear, but in the itineraries that these photos have followed online. And however we would like to qualify their cohesion if not coherence, we should not look for a panorama, even incomplete and fragmented, but for other modes of stitching and splitting, of combining their storage volumes and conjunctive surfaces.

Notes

1. ↑ Phil Askey, "Kodak DC280 Review," DPReview, accessed November 19, 2020.

2. ↑ Askey, "Kodak DC280 Review," 6.

3. ↑ In an article related to the dataset Caltech 256, Pietro Perona, pictured in *faces1999*, uses the word *clutter* to describe what is not the main object of the photo. G. Griffin, A. Holub, and P. Perona, "Caltech-256 Object Category Dataset," CalTech Report, 2007.

4. ↑ Geoff Cox, Nicolas Malevé and Michael Murtaugh, "Archiving the Data Body: Human and Nonhuman Agency in the Documents of Erkki Kurenniemi," in *Writing and UnWriting Media (Art) History: Erkki*

Kurenniemi in 2048, eds. Joasia Krysa and Jussi Parikka (Cambridge MA; MIT Press, 2015), 125-142.

5. ↑ Ankan Bansal et al., "UMDFaces: An Annotated Face Dataset for Training Deep Networks," *CoRR,* (2016), abs/1611.01484. Available from: http://arxiv.org/abs/1611.01484.

6. ↑ Qiong Coa et al., "VGGFace2: {A} dataset for recognising faces across pose and age," *CoRR,* (2017), abs/1710.08092. Available from: http://arxiv.org/abs/1710.08092.

7. ↑ For an example of such a narrative, see Andrey Kurenkov, "A 'Brief' History of Neural Nets and Deep Learning, Part 1," accessed January 2, 2017, http://www.andreykurenkov.com/writing/a-brief-history-of-neural-nets-and-deep-learning.

8. ↑ On the soft versus hard rule, see Lorraine Daston, "Algorithms Before Computers," Simpson Center for the Humanities UW (2017), accessed March 25, 2020, https://www.youtube.com/watch?v=pqoSMWnWTwA.

9. ↑ Karen Barad, "Meeting the Universe Halfway: Realism and Social Constructivism without Contradiction," in Nelson, L. H. and Nelson, J. (eds.) *Feminism, Science, and the Philosophy of Science.* Dordrecht: Springer Netherlands, 1996, 161-194.

10. ↑ Vilem Flusser, *Towards a Philosophy of Photography* (London: Reaktion books, 2000).

The Fragility of Life

Simone C Niquille in conversation with Jara Rocha
and Femke Snelting

This text was edited from a conversation, recorded after the screening of process material for Niquille's film *The Fragility of Life* at the Possible Bodies residency in Akademie Schloss Solitude, Stuttgart (May 2017).

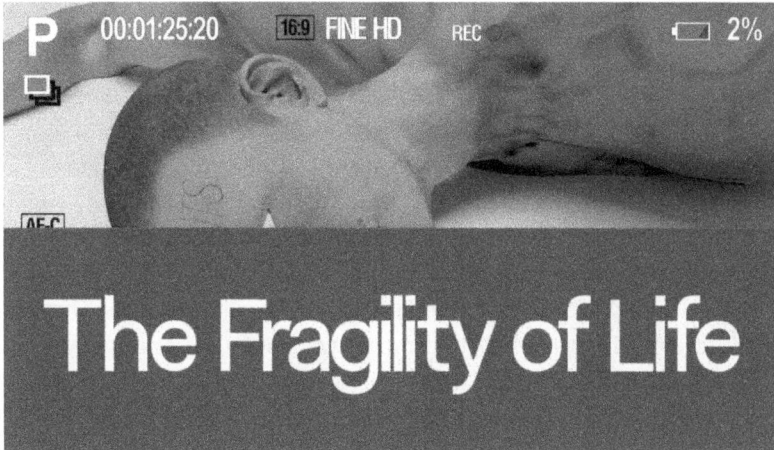

Simone C Niquille, *The Fragility of Life*, 2017, filmstill

Jara Rocha: In the process of developing the Possible Bodies trajectory, one of the excursions we made was to the Royal Belgian Institute of Natural Science's reproduction workshop in Brussels, where they were working on 3D-reproductions of Hominids. Another visitor asked: "How do you know how many hairs a monkey like this should have?" The person working on the 3D reproduction replied, "It is not a monkey."[1] You could see that he had an empathetic connection to the on-screen-model he was working on, being of the same species. I would like to ask you about norms and embedded norms in software. Talking about objective truth and parametric representation and the like, in this example you refer to, there is a huge norm that worries me, that of species, of unquestioned humanness. When we talk about "bodies", we can push certain limits because of the hegemony of the species. In legal court, the norm is anthropocentric, but when it comes to representation...

Femke Snelting: This is the subject of "Kritios They"?

Simone C Niquille: Kritios They is a character in *The Fragility of Life*, a result of the research project *The Contents*. While *The Contents* is based on the assumption that we as humans possess and create content, living in our daily networked space of appearance that is used for or against us, I became interested in the corporeal fragility exposed and created through this data, or that the data itself possesses. In the film, the decimation scene questions this quite bluntly: when does a form stop being human, when do we lose empathy towards the representation? Merely reducing the 3D mesh's resolution, decreasing its information density, can affect the viewer's empathy. Suddenly the mesh might no longer be perceived as human, and is revealed as a simple geometric construct: A plain surface onto which any and all interpretation can be projected. The contemporary accelerating frenzy of collecting as much data as possible on one single individual to achieve maximum transparency and construct a "fleshed out" profile is a fragile endeavor. More information does not necessarily lead to a more defined image. In the case of Kritios They, I was interested in character creation software and the parameters embedded in its interfaces. The parameters come with limitations: an arm can only be this long, skin color is represented within a specified spectrum, and so on. How were these decisions made and these parameters determined? Looking at design history and the field's striving to create a standardized body to better cater to the human form, I found similarities of intent and problematics.

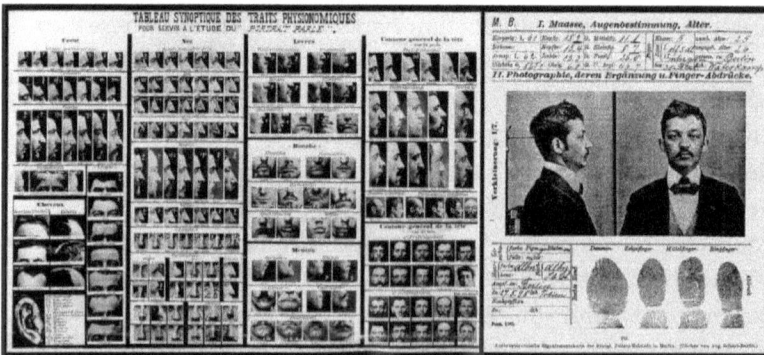

Alphonse Bertillon, Anthropometric data sheet and Identification Card, 1896

Anthropometric efforts ranging from Da Vinci's *Vitruvian Man*, to Corbusier's *Modulor*, to Alphonse Bertillon's *Signaletic Instructions* and invention of the mug shot, to Henry Dreyfuss's *Humanscale*... What these projects share is an attempt to translate the human body into numbers. Be it for the sake of comparison, efficiency, policing...

In a Washington Post article from 1999[2] on newly developed voice mimicking technology, Daniel T. Kuehl, the chairman of the Information Operations department at the National Defense University in Washington (the military's school for information warfare) is quoted as saying: "Once you can take any kind of information and reduce it into ones and zeroes, you can do some pretty interesting things."

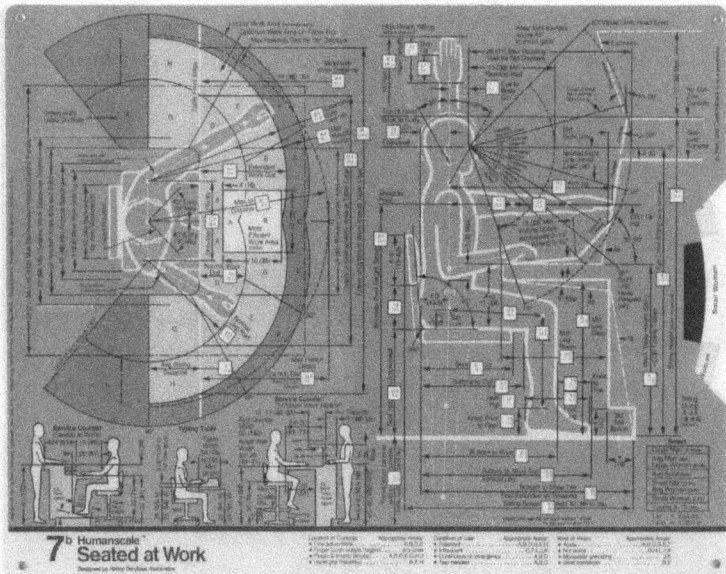

Humanscale 7b: Seated at Work Selector, Henry Dreyfuss Associates, MIT Press, 1981. Photo: Courtesy of Cooper Hewitt, Smithsonian Design Museum http://collection.cooperhewitt.org/objects/51689299

To create the "Kritios They" character I used a program called Fuse.[3] It was recently acquired by Adobe and is in the process of being integrated into their Creative Cloud services. It originated as assembly-based 3D modeling research carried out at Stanford University. The Fuse interface segments the body into Frankenstein-like parts to be assembled by the user. However, the seemingly restriction free Lego-character-design interface is littered with limitations. Not all body parts mix as well as others; some create uncanny folds and seams when assembled. The torso has to be a certain length and the legs positioned in a certain way and when I try to adapt these elements the automatic rigging process doesn't work because the mesh won't be recognized as a body.

A lot of these processes and workflows demand content that is very specific to their definition of the human form in order to function. As a result, they don't account for anything that diverges from that norm, establishing a parametric truth that is biased and discriminatory. This raises the question of what that norm is and how, by whom and for whom it has been defined.

FS: Could you say something about the notion of "parametric truth" that you use?

SN: Realizing the existence of a built-in anthropometric standard in such software, I started looking at use cases of motion capture and 3D scanning in areas other than entertainment — applications that demand an objectivity. I was particularly interested in crime and accident reconstruction animations that are produced as visual evidence or in court support material. Traditionally this support material would consist of photographs, diagrams and objects. More recently this sometimes includes forensic animations commissioned by either party. The animations are produced with various software and tools, sometimes including motion capture and/or 3D scanning technologies.

These animations are created post-fact; a varying amalgam of witness testimonies, crime scene survey data, police and medical reports etc. Effectively creating a "version of", rather than an objective illustration. One highly problematic instance was an animation intended as a piece of evidence in the trial of George Zimmerman on the charge of second-degree murder on account of the shooting of Trayvon Martin in 2012. Zimmerman's defense commissioned an animation to attest his actions as self defense. Among the online documentation of the trial is a roughly two-hour long video of Zimmerman's attorney questioning the animator on his process. Within these two hours of questioning, the defense attorney is attempting to demonstrate the animations' objectivity by minutely scrutinizing the creation process. It is revealed that a motion capture suit was used to capture the character's animations, to digitally re-enact Zimmerman and Martin. The animator states that he was the one wearing the motion capture suit portraying both Zimmerman as well as Martin. If this weren't already enough to debunk an objectivity claim, the attorney asks: "How does the computer know that it is recording a body?" Upon which the animator responds: "You place the sixteen sensors on the body and then on screen you see the body move in accordance." But what is on screen is merely a representation of the data transmitted by 16 sensors, not a body.

A misplaced or wrongly calibrated sensor would yield an entirely different animation. And further, the anthropometric measurements of the two subjects were added in post production, after the animation data had been recorded from the animator's re-enactment. In this case the animation was thankfully not allowed as a piece of evidence, but it nevertheless was allowed to be screened during the trial. The difference from showing video in court is, seeing something play out visually, in a medium that we are used to consume. It takes root in a different part of your memory than a verbal acount and renders one version more visible than others. Even with part of the animation based on data collected at the crime scene, a part of the reproduction will remain approximation and assumption.

3D animation by Reuter's owned News Direct "Transform your News with 3D Graphics", "FBI investigates George Zimmerman for shooting of Florida teen, Trayvon Martin", *News Direct*, 2012

This is visible in the visual choices of the animation, for example. Most parts are modeled with minimal detail (I assume to communicate objectivity). "There were no superfluous aesthetic choices made." However, some elements receive very selective and intentional detailing. The crime scene's grassy ground is depicted as a flat plane with an added photographic texture of grass rather than 3D grass produced with particle hair. On the other hand, Zimmerman and Martin's skin color is clearly accentuated as well as the hoodie worn by Trayvon Martin, a crucial piece of the defense's case. The hoodie was instrumentalized as evidence of violent intentions during the trial, where it was claimed that if Martin had not worn the hood up he would not have been perceived as a threat by Zimmerman. To model these elements at varying subjective resolution was a deliberate choice. It could have depicted raw armatures instead of textured figures, for example. The animation was designed to focus on specific elements; shifting that focus would produce differing versions.

FS: This is something that fascinates me, the different levels of detailing that occur in the high octane world of 3D. Where some elements receive an enormous amount of attention and other elements, such as the skeleton or the genitals, almost none.

SN: Yes, like the sixteen sensors representing a body...

FS: Where do you locate these different levels of resolution?

SN: Within the CGI [computer-generated imagery] community, modelers are obsessed by creating 3D renders in the highest possible resolution as a technical as well as artistic accomplishment, but also as a form of muscle flexing of computing power. Detail is not merely a question of the render quality, but equally importantly it can be the realism achieved; a tear on a cheek, a thin film of sweat on the skin. On forums you come across discussions on something called subsurface scattering,[4] which is used to simulate blood vessels under the skin to make it look more realistic, to add weight and life to the hollow 3D mesh. However, the discussions tend to focus on pristine young white skin, oblivious to diversity.

JR: This raises the notion of the "epistemic object". The matter you manipulated brings a question to a specific table, but it cannot be on every table: it cannot be on the "techies" table *and* on the designers table. However, under certain conditions, with a specific language and political agenda and so on, *The Contents* raises certain issues and serves as a starting point for a conversation or facilitates an argument for a conversation. This is where I find your work extremely interesting. I consider what you make objects around which to formulate a thought, for thinking about specific crossroads. They can as such be considered a part of "disobedient action-research", as epistemic objects in the sense that they make me think, help me wonder about political urgencies, techno-ecological systems and the decisions that went into them.

SN: That's specifically what two scenes in the film experiment with: the sleeping shadow and the decimating mug shot. They depend on the viewer's expectations. The most beautiful reaction to the decimating mug shot scene has been: "Why does it suddenly look so scary?"

The viewer has an expectation in the image that is slowly taken away, quite literally, by lowering the resolution. Similar with the sleeping scene: What appears as a sleeping figure filmed through frosted glass unveils itself by changing the camera angle. The new perspective reveals another reality. What I am trying to figure out now is how the images operate in different spaces. Probably there isn't one single application, but they can be in *The Fragility of Life* as well as in a music video or an ergonomic simulation, for example, and travel through different media and contexts. I am interested in how these images exist in these different spaces.

FS: We see that these renderings, not only yours but in general, are very volatile in their ability to transgress applications, on the large scale of movements ranging from Hollywood to medical, to gaming, to military. But it seems that, seeing your work, this transgression can also function on different levels.

SN: These different industries share software and tools, which are after all developed within their crossroads. Creating images that attempt to transgress levels of application is a way for me to reverse the tangent, and question the tools of production.

Is the image produced differently if the tool is the same or is its application different? If 3D modeling software created by the gaming industry were used to create forensic animations, possibly incarcerating people, what are the parameters under which that software operates? This is a vital question affecting real lives.

JR: Can you please introduce us to Mr. #0082a?

SN: In attempting to find answers to some of the questions on the Fuse character creator software's parameters I came across a research project initiated by the U.S. Air Force Research Laboratory from the late 1990s and early 2000s called "CAESAR" [Civilian American and European Surface Anthropometry Resource].

#0082a is a whole body scan mesh from the CAESAR database,[5] presumably the 82nd scanned subject in position a. The CAESAR project's aim was to create a new anthropometric surface database of body measurements for the Air Force's cockpit and uniform design. The new database was necessary to represent the contemporary U.S. military staff. Previous measurements were outdated as the U.S. population had grown more diverse since the last measurement standards had been registered. This large-scale project consisted of scanning about 2000 bodies in the United States, Italy and the Netherlands. A dedicated team travelled to various cities within these countries outfitted with the first whole body scanner developed specifically for this purpose by a company called Cyberware. This is how I initially found out about the CAESAR database, by trying to find information on the Cyberware scanner.

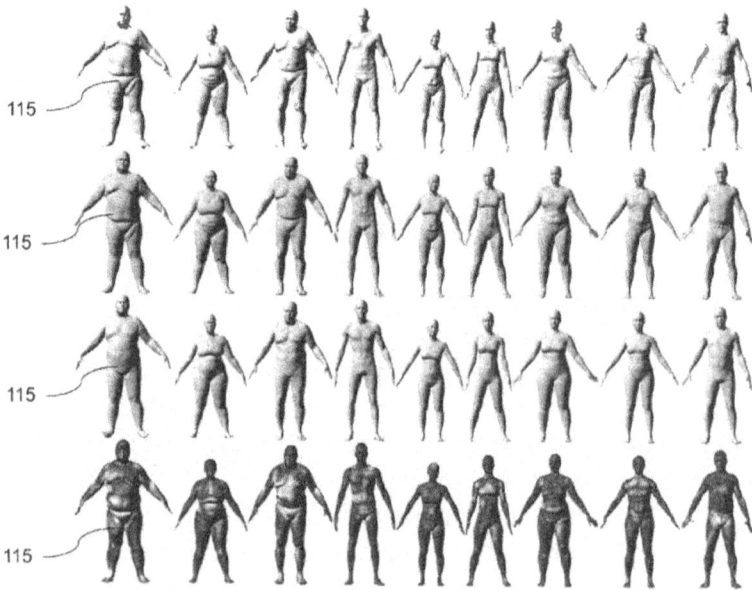

CAESAR database used as training set in the research towards a parametric three-dimensional body model for animation. Loper et al, "Method for providing a threedimensional body model," patent US 10,417,818 B2 filed by Max-Planck-Gesellschaft zur Förderung der Wissenschaften e.V., 2019

I found a video somewhere deep within YouTube, it was this very strange and wonderful video of a 3D figure dancing on a NIST [U.S. National Institute of Standards and Technology] logo. The figure looked like an early 3D scan that had been crudely animated. I got in touch with the YouTube user and through a Skype conversation learned about his involvement in the CAESAR project through his work at NIST. Because of his own personal fascination with 3D animation he made the video I initially found by animating one of the CAESAR scans, #0082a, with an early version of Poser.

Cyberware[6] has its origins in the entertainment industry. They scanned Leonard Nimoy, who portrays Spock in the Star Trek series, for the famous dream sequence in the 1986 movie Star Trek IV: The Voyage Home. Nimoy's head scan is among the first 3D scans... The trajectory of the Cyberware company is part of a curious pattern: it originated in Hollywood as a head scanner, advanced to a whole body scanner for the military, and completed the entertainment-military-industrial cycle by returning to the entertainment industry for whole-body scanning applications.

CAESAR, as far as I know, is one of the biggest databases available of scanned body meshes and anthropometric data to this day. I assume, therefore it keeps on being used — recycled — for research in need of humanoid 3D meshes.

While looking into the history of the character creator software Fuse I sifted through 3D mesh segmentation research, which later informed the assembly modeling research at Stanford that became Fuse. #0082 was among twenty CAESAR scans used in a database assembled specifically for this segmentation research and thus ultimately played a role in setting the parameters for Fuse. A very limited amount of training data, that in the case of Fuse ended up becoming a widely distributed commercial software. At least at this point the training data should be reviewed... It felt like a whole ecology of past and future 3D anthropometric standards revealed itself through this one mesh.

Notes

1. ↑ See also: "We hardly encounter anything that didn't matter," in this book.
2. ↑ William M. Arkin, "When Seeing and Hearing Isn't Believing," *Washington Post*, February, 1999, https://www.washingtonpost.com/gdpr-consent/?next_url=https%3a%2f%2fwww.washingtonpost.com%2fwp-srv%2fnational%2fdotmil%2farkin020199.htm.
3. ↑ Jeanette Mathews, "An Update on Adobe Fuse as Adobe Moves to the Future of 3D & AR Development," September 13, 2019, https://www.adobe.com/products/fuse.html.
4. ↑ "Subsurface Scattering," *Blender 2.93 Reference Manual*, accessed July 1, 2020, https://docs.blender.org/manual/en/latest/render/shader_nodes/shader/sss.html.
5. ↑ Products based on this database are commercialized by SAE International, http://store.sae.org.
6. ↑ "Cyberware," Wikipedia, accessed July 1, 2020, https://en.wikipedia.org/wiki/Cyberware.

Rehearsal as the 'Other' to Hypercomputation

Maria Dada

The next few paragraphs outline the effects of the simulation paradigm on the sense of errantry in the postcolonial condition in places like Lebanon. Through an examination of two games about the Beirut war, that differ in their approach, the text examines the possibilities of opening up a space for the Other in the gap between simulation as rehearsal versus that of training.

History is apparently no longer sufficient to uphold the dominance of the western viewpoint. It must be overcome, but despite the prevalence of critical tools such as discourse analysis, genealogy archaeology and other methods that attempt to dismantle the totalitarian universal structure of history, it is simulation that appears to disassemble it, only to take its place. However, to overcome history through simulation is to root the colonized into a past of prediction, efficiency and closed repetition. Simulation studies people and places like Beirut and their wars as strategy, in order to lock them into a position that is not indigenous to their way of being, that of errantry.

Édouard Glissant describes errantry as "rooted movement" in a sense that it's a desire to go against the root where the root is the historical beginnings and universalization of the western point of view.[1] The history of the West has always been tied to fixed states of nationality, an idea that has been exported to the colonized nations like Lebanon, that have come to aspire to similar univocal rootedness. The idea of errantry, which Glissant believes to be native to the colonized, is a fluid subjectivity that sits between the notion of identity and movement.

In other words, what this text will put forward, is that simulation closes in on the possibilities of what Glissant describes as a poetics, creating a continuous longing for the lost but defunct and deconstructed stories of origins and history that are tied to the west. In order to foreclose on the pasts war like the one of Beirut in 1982, simulation engines use remote sensing and computer-generated images to build model worlds in order to programmatically train on different scenarios, from different perspectives across different surfaces of the earth. Simulation becomes a device to train actions and access history in a world of greater perceived uncertainly, automation, deregulation and the supposed "need" for risk management.

I need not reiterate the pages and pages written on the prominence of economics-based calculation and prediction of events that have taken over from poetry, storytelling and meaning; the decreasing importance of a stable and single point of view which is being supplemented (and often replaced) by multiple perspectives, overlapping windows, distorted flight lines, and divergent vanishing points. Farewell to History which should have long been replaced by genealogy, archaeology, discourse analysis and the evolutionary vibrations of matter, geology and exploding events, exploding long before history, contingency and accidents bubbling beneath the crust. A loss that is felt even more prominently these days with the constant interruption of screen face-to-face conversations by glitches, echoes, ventilation hum, or simply by headaches and sore eyes.

The representational scalar vocabularies of narrative storytelling are no longer good enough to describe the complex temporality and spatiality of the world. One that appears to be a composite matter of deep time water undersea, rocks, stones, forests, the body feminine, the marginalized, the repressed, the unconscious, and the algorithms. Global infrastructures, computer generated images, data behaviorism, all of the aspects of the new geo-political and economic interdependencies that make up our world. Simulation and tactical gameplay have come to replace historical folktales. History as a fictional linear progression that continuously follows on from event to event, that has a form of unity, western rootedness and continuity, is no longer perceived as sufficient enough to describe the diverse multitude of our current reality. History is the discourse of the powerful; it's the discourse of totalitarianism, of hegemony, which must be critiqued and questioned. Unfortunately these critical methods forever remain buried behind the thrust and efficacy of modeling volumetric unknowns.

Simulation is the new method of certainty, which borrows its art from cybernetics, particle physics and statistical mechanics. It has come to replace history, to break up its hold on reality, by presenting the past through multiplying perspectives. However, simulation comes in two flavors, that of *rehearsal* and of *training*. The latter is always seemingly co-opted and incorporated into volumetric regimes of the probable closing in on all the possibilities that could open up when history dissolves. The chaotic weather systems of social, political, animate and post-colonial perspectives, under the current regimes of volumetric terror, or simulation as training are tamed, suffocated in predictive echo chambers, from contingency and accident to calculated probabilities.

However, not all tactical gameplay is designed the same, simulations can appear as rehearsal on the one hand and training on the other. An example of simulation-as-training as opposed to rehearsal is the way that crewless vessels or autonomous cargo ships are trained through various volumetric exercises and modeling. To understand the possibilities of traversing the sea in the shortest amount of time, with the least amount of trouble. The experiments of the sea of past and beyond are no longer there. An autonomous ship does not sail for exploration, however problematic that term is, when considering colonial encounters. Even the colonial ships that wanted to discover and conquer, left a little bit of space for contingency, for the accident. The cargo autonomous ships, however, leave no room or margin for error. They must train for all scenarios regardless of their position. And if these ships encounter a scenario that is not part of the training package, then they no longer know what to do. A failure in this sense is not an opportunity for discovery, a failure is complete deadlock. The training of volumetric regimes is a future speculative exercise for closing up the future for minimizing error and risk. Furthermore, the difference between the rehearsal of the first-person taking command of the simulation engine on the one hand and the rehearsal of the autonomous machine learning system that is acting as an opponent on the other, is that simulators mould, through training, the corpus of living beings to the machine, while the autonomous system extracts the bodily presence from the rehearsal process. It's not training the body anymore it's training of data archived, extracted.

With the number of simulations trialed at the moment, it's almost as if we've entered some form of "Training Paradigm", that is if we could ever again believe in the phenomena of paradigms or epochs. From marketing campaigns to political campaigns training on consumer or voter temperament, to competing models simulating virus paths, vaccine efficacy and the rate at which black and ethnics minorities are likely to get infected due to frontline jobs they are forced into by structural racism. Train the timeline, train for the unlikely scenario, learn the drill and prepare for the victor. Prepare the seven speeches only to read out the one that seems most fitting when you know the results. To train, a preparation for pointless antic-ipated activities.

The term "re-hearse" combines the Latin (re) with the old French *herse*, meaning harrow or a large rake used to turn the earth or ground, as in to reground or to take the ground again, to rake it again until all possible grounds have been considered.[2] A distinction, how-ever, should be made between rehearsal and training. If rehearsal is the repetition that maintains the openness of the rehearsed piece, a repetition that produces difference each time the piece is rehearsed, then training is the moment of closure in the process of rehearsal, when contingency is purposefully erased. What training does as every performer knows is that it destroys the spontaneity of the moment, "The performer, therefore, could not rehearse such music but rather "trained" for it like a martial art, developing ways of acting upon con-tingency."[3]

Training is in this sense different from the practice of rehearsal, which is a gesture of putting something into action, from the theory into practice. To train for something is to consider and attempt to foreclose all possible futures by unearthing various possible grounds for any future. When one trains they repeat an action in an attempt to erase the possibility for the accident, or erase the possibility of any kind of error. Training for a sport, for example, tends to optimize all the muscles towards a very specific and closed, aim that leaves no room for the accident. The accident in sport is always an injury.

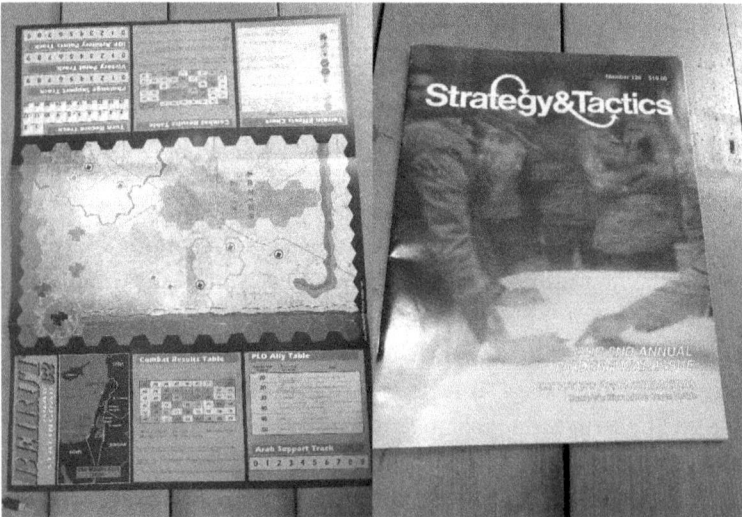

GAME RULES PAGE 1: On June 13th, 1982, paratroopers and armour of the Israeli Defence Forces (IDF) rolled to the edge of Beirut, joining forces with their Phalange Christian allies. They never got much farther. The Palestine Liberation Organisation's attempt to organises a regular army had failed, but so had Israel's drive to exterminate it. This was a classic confrontation of modern diplomacy, where political pressure allowed a tiny force to fend off a giant. Beirut '82: Arab Stalingrad simulates the siege of Beirut, and its victory conditions recreate the diplomatic hindrances of that struggle.[4]

The above excerpt is taken from the 1989 edition of *Strategy and Tactics* magazine which was founded in 1966 by a US Air Force Staff Sergeant named Chris Wagner. The point of the magazine, or "war fanzine", was to produce more complex and therefore more realistic tactics in wargaming. The magazine had elements of a recreational wargaming magazine but as it was written by military political analysts and defense consultants who were keen to create something close enough to military wargaming. In 1969 James F. Dunnigan, a political analyst, formed Simulations Publications, Inc., a publishing house created specifically to publish the magazine.[5]

The excerpt is the first paragraph of the game rule page that explains the rules of *Beirut '82: Arab Stalingrad*, a game based on *The Siege of Beirut*, one of the most defining events of the Lebanese Civil War. The siege took place in the summer of 1982 when the United Nation ceasefire between the Palestinian Liberation Army (PLO), who in the early 1970s made Lebanon its base of operations, and the Israeli army. After the siege the PLO were forced out of Beirut and the rest of Lebanon. *Strategy and Tactics* was one of the first wargaming magazines to include a wargame within its pages.

The main difference between so-called recreational wargames such as *Beirut '82: Arab Stalingrad*, however realistic and complex they intend to be, and military wargames, is that the former is usually regarded as a historical depiction of war. The training on tactics and strategies is replaying the events of a distant past. Wargaming has long performed World Wars I and II and the Napoleonic Wars as an act of remembrance and an interest of historians. Recreational games generally take creative liberties, by adding fictional elements, to make the game more enjoyable, more playable. For instance, scenarios would often be differently simplified in order to prioritize gameplay over event accuracy. However, *Strategy and Tactics* as a magazine that sits between tactical history and military strategy prides itself on being more realistic than other wargaming magazines.

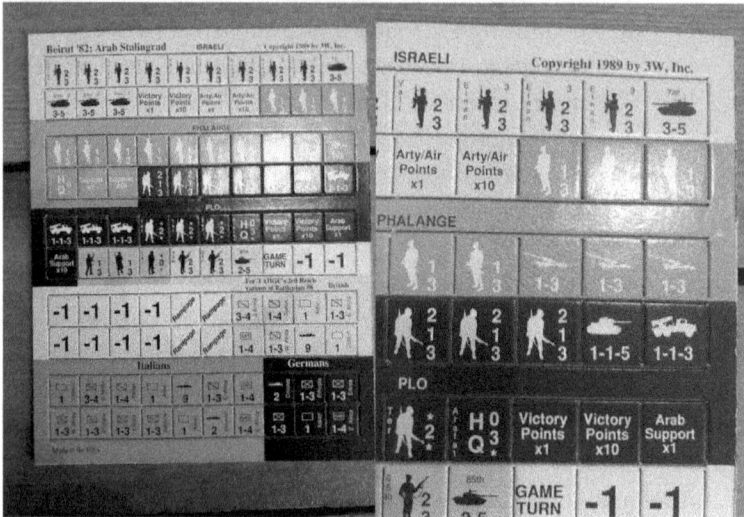

GAME RULES PAGE 5, 6.0 CIVILIAN CASUALITIES: The CRT (rule 4.22) shows if an attack might cause Civilian Casualties, and what to multiply the result by. However, these casualties still only occur under certain conditions. IDF units or artillery points must participate in the attack and the PLO must be defending a Refugee Camp or City hexagon. Otherwise, ignore Civilian Casualties.[6]

Wargaming is a descriptive and predictive apparatus that goes beyond the magazines and technologies of its implementation. When playing a game such as *Beirut '82: Arab Stalingrad* on the map insert placed in the centerfold of the publication, the gamer moves the Phalange army troops, as cardboard cut-outs of a right-wing Maronite party in Lebanon founded in 1936 by Pierre Gemayel, across the map. Such a move is a re-enactment of a particular procedure that relates to a complex system which reproduces what to some are painful historical events in relation to other possible futures, possible or probable futures that will never be. The combat is replaced with abstraction, supply and demand dynamics and other military considerations of algorithmic and numerically founded sets of possible outcomes all made random, a flipping of events at the throw of a die. *Beirut '82: Arab Stalingrad* is interesting not for its own sake but for in the manner in which it represents knowledge or history as a combination of both rehearsal and training, as simulation, or as Haron Farocki describes, "life trained as a sport".[7]*Beirut '82: Arab Stalingrad* is not only a simulation: it is one of the most nuanced and complex examples found in any medium. As a game it has eight pages of rules which explain actions, moves and procedures for circa one hundred game pieces and tokens around a 50cm by 40cm battle ground map of Beirut. It allows for a physically as well as conceptually extreme level of gameplay.

More than this, software gaming, from its inception, was quick to take interest in wargaming, which is different from games with military themes. Wargames were quick to translate to the screen and themes of Beirut 82 were no exception. The difference being that simulation now attempts to model all of the weapons, vehicles and aircrafts that were involved in the siege for show. Digital Combat Simulator's UH-1H Huey mission entitled Beirut 82 is an exemplar of the wargaming simulation offering a first-person experience of what it's like to be an American built Israeli helicopter flying over Beirut in 1982. The DCS website describes it as:

> *Digital Combat Simulator World (DCS World) 2.5 is a free-to-play digital battlefield game. Our dream is to offer the most authentic and realistic simulation of military aircraft, tanks, ground vehicles and ships possible… DCS: UH-1H Huey features an incredible level of modelling depth that reproducers the look, feel, and sound of this legendary helicopter with exquisite detail and accuracy. Developed in close partnership with actual UH-1H operators and experts, the DCS Huey provides the most dynamic and true to life conventional helicopter experience available on the PC. The UH-1 Huey is one of the most iconic and recognisable helicopters in the world. Having served extensively as a transport and armed combat support helicopter in the Vietnam War, the Huey continues to perform a wide variety of military and civilian missions around the world today.[8]*

Here the simulation is less interested in the historical strategies that play out a future otherwise. The volumetrics of the UH-1H Huey are there to both produce a so-called "modeling depth" in order to train the gamer to fly the helicopter over the terrain Beirut. The "modeling depth" of the Huey relates to a calculated time of the clock, not a temporality of sorts, but rather a time in the milliseconds, for instance, that it takes to fly the aircraft for calculations sake: calculation for calculations sake.

"Modeling depth" also relates to the attention to visual and volumetric detail in the construction of the aircraft itself, to resolution. Depth here considers only pixel resolution of the type of visual dimension that captures the aircraft in a hyper computational state. It means very little to the people on the ground, viewing it as it shells its missiles, or captures prisoners who are their family members on the ground. In effect none of the events of Beirut 82 are captured in this simulation, not even the tactics and facts of history.

So, while in *Strategy and Tactics*, and the Beirut 82 replaying there is the probable and possible future that can be played and played again, even if it will never be realized. There is an opening for discussion of the past. In that sense, the past is being rehearsed as if it could have been otherwise. The tactical re-playing of past in that sense becomes a mode of open discussion within the game. Historical recollection can no longer be a simple story, narrative or folklore. Historical recollection has to include tactical exercises, a replaying, a repetition, a habit, a form of inhabiting the past which keeps its own tactical memories; the memory or schema of a victory that's played as tactical exercise. With the DCS: UH-1H Huey, however, the body of the gamer trains to fly the helicopter where the training is performed at the individual level siloed in the aircraft shooting down at the landscape, practicing nothing but flight skills and good aim.

Only in the openings between the tactical gameplay can there be remnants or conversations about an "Other" to hypercomputation. The slight opening in *Beirut '82: Arab Stalingrad*, within the gameplay allows for a possible outside of the probable. However, it does so in a manner that never gets cemented into writing, into a root that can be acted against, in a manner that aligns with what Glissant defined as "errantry". We do not yet know the general movement of errantry, "the desire to go against the root", the indigenous being of the colonized in relation to simulation, whether training or rehearsal.[9] Even if *Beirut '82: Arab Stalingrad* permits a type of gameplay it remains a pointlessly ephemeral, fleeting moment that passes away as the game ends but leaves behind nothing but loss. It leaves the question open; How can the "training" convert into "rehearsal"?

Notes

1. ↑ Édouard Glissant. *Poetics of Relation* (University of Michigan Press, 1997).

2. ↑ * Rehearse. (n.d.). Retrieved from https://www.merriam-webster.com/dictionary/rehearse

3. ↑ * Simon Yuill, "All Problems of Notation Will Be Solved By the Masses," in *Mute* Vol 2, No. 8, 2008 https://www.metamute.org/editorial/articles/all-problems-notation-will-be-solved-masses.

4. ↑ "Beirut '82: Arab Stalingrad," *Strategy and Tactics*, 1989.

5. ↑ S. Appelcline, *Designers & dragons*. Silver Springs, MD: Evil Hat Productions, 2013.

6. ↑ "Beirut '82: Arab Stalingrad," Strategy and Tactics.

7. ↑ Thomas Elsaesser, "Simulation and the Labour of Invisibility: Harun Farocki's Life Manuals," *Animation* 12, no. 3 (November 2017): 214–29.

8. ↑ *Digital Combat Simulator World* (n.d.), retrieved from https://www.digitalcombatsimulator.com/en/products/world/.

9. ↑ Glissant, *Poetics of Relation*.

We hardly encounter anything that didn't really matter

Phil Langley in conversation with Possible Bodies

As an architect and computational designer, Phil Langley develops critical approaches to technology and software for architectural practice and spatial design. Our first conversation started from a shared inquiry into MakeHuman,[1] the Open Source software project for modeling 3-dimensional humanoid characters.

In the margins of the yearly Libre Graphics meeting in Toronto, we spoke about the way that materiality gets encoded into software, about parametric versus generative approaches, and the symbiotic relationship between algorithms that run simulations and the structure of that algorithm itself. "I think there is a blindness in understanding that the nature of the algorithm effects the nature of the model... The model that you see on your screen is not the model that is actually analyzed."[2]

Six years later, we ask him about his work for the London-based architecture and engineering firm Bryden Woods where he is now responsible for a team that might handle computational design in quite a different way.

A very small ecosystem

Phil Langley: For the Creative Technologies team that I set up in my company, we hired twenty people doing computational design and they all come from very similar backgrounds: architectural engineering plus a postgraduate or a master's degree in "computational design". We all have similar skills and are from a narrow selection of academic institutions. It is a very small ecosystem.

I followed a course around 2007 that is similar to what people do now. There's some of the technology that moves on for sure, but you're still learning the same kind of algorithms that were there in the 1950s or sixties or seventies. They were already old when I was doing them. You're still learning some parametrics, some generative design, generative algorithms, genetic algorithms, neural networks and cellular automatisms, it is absolutely a classic curriculum. Same texts, same books, same references. A real echo chamber.

One of the things I hated when I studied was the lack of diversity of thoughts, of criticality around these topics. And also the fact that

143

there's only a very narrow cross-section of society involved in creating these kinds of techniques. If you ever mentioned the fact that some of these algorithmic approaches came from military research, the response was: "So what?". It wasn't even that they said that they already knew that. They were just like "Nothing to say about that, how can that possibly be relevant?"

How can you say it actually works?

PL: When building the team, I was very conscious about not stepping straight into the use of generative design technologies, because we certainly haven't matured enough to start the conversation about how careful you have to be when using those techniques. We are working with quite complex situations and so we can't have a complex algorithm yet because we have too much to understand about the problem itself.

We started with a much more parametric and procedural design approach, that was much more... I wouldn't say basic... but lots of people in the team got quite frustrated at the beginning because they said, we *can* use this technique, why don't we just use this? It's only this year that we started using any kind of generative design algorithms at all. It was forced on us actually, by some external pressures. Some clients demanded it because it becomes very fashionable and they insisted that we did it. The challenges or the problems or the kind of slippage is how to try and build something that uses those techniques, but to do it consciously. And we are not always successful achieving that, by the way.

The biggest thing we were able to achieve is the transparency of the process because normally everything that you pile up to build one of those systems, gets lost. Because it is always about the performance of it, that is what everybody wants to show. They don't want to tell you how they built it up bit by bit. People just want to show a neural network doing something really cool, and they don't really want to tell you how they encoded all of the logic and how they selected the data. There are just thousands of decisions to make all the way through about what you include, what you don't include, how you privilege things and not privilege other things.

At some point, you carefully smooth all of the elements or you de-noise that process so much... You simplify the rules and you simplify the input context, you simplify everything to make it work, and then how can you say that it actually works? Just because it executes and doesn't crash, is that really the definition of functionality, what sort of truth does it tell you? What answers does it give you?

You make people try to understand what it does and you make people talk about it, to be explicit about each of those choices they make, all those rules, inputs, logics, geometry or data, what they do to turn that into a system. Every one of those decisions you make defines the n-dimensional space of possibilities. And if you take some very complicated input and you can't handle it in your process and you simplify so much, you've already given a shape to what it could possibly emerge as. So one of the things we ended up doing is spending a lot of time on that and we discuss each micro step. Why are we doing it like this? It wasn't always easy for everyone because they didn't want to think about documenting all the steps.

Yesterday we had a two hour conversation about mesh interpolation and the start of the conversation was a data flow diagram, and one of the boxes just said something like: "We're just going to press this button and then it turns into a mesh". And I said: "Woah, wait a minute!" some people thought "What do you mean, it's just a feature, it's just an algorithm. It's just in the software, we can just do it." And I said, "No way." That's even before you get towards building something that acts on that model. I think that's what we got out of it actually, by not starting with the most let's say sophisticated approach, it has allowed us to have more time to reflect on what fueled the process.

Decisions have to be made

Possible Bodies: Do you think that transparency can produce a kind of control? Or that 'understanding' is somehow possible?

PL: It depends what you mean by control, I would say.

It is not necessarily that you do this in order to increase the efficacy of the process or to ensure you get better results. You don't do it in order to understand all of the interactions because you cannot do that, not really. You can have a simpler algorithmic process, you can have an idea of how it's operating, there is some truth in that, in the transparency, but you lose that quite quickly as the complexity grows, it's more to say that you re-balance the idea that you want to see an outcome you like, and therefore then claim that it works. I want to be able to be explicit about everything that I know all the way long. In the end that's all you have. By making explicit that you have made all these steps, you make clear that decisions have to be made. That at every point you're intervening in something, and it will have an effect.

Almost every one of these things has an effect to a greater or lesser extent and we hardly encounter anything that didn't really matter. Not even if it was a bug. If it wasn't really affecting the system, it's probably because it was a bug in the process rather than anything else.

I think that transparency is not about gaining control of a process in itself, it's about being honest with the fact that you're creating something with a generative adversarial network (GAN) or a neural network, whatever it is. That it doesn't just come from TensorFlow,[3] fully made and put into your hand and you just press play.

Getting lost in nice little problems

PL: The point I was trying to make to everyone on the team was, well, if you simplify the mesh so much in order that it's smooth and so you can handle it in the next process, what kind of reliance can you have on the output?

I'll tell you about a project that's sort of quite boring. We are developing an automated process for cable rooting for signaling systems in tunnels. We basically take a point cloud survey of a tunnel and we're trying to route this cable between obstacles. The tunnel is very small, there is no space, and obviously there's already a signaling system there. So there are cables everywhere and you can't take them out while you install the new ones, you have to find a pathway. Normally this would be done manually. Overnight people would go down in the tunnel and spray paint the wall and then photograph it and then come back to the office and try and draw it. So we're trying do this digitally and automate it in some way. There's some engineering rules of the cables, that have to be a certain diameter. You can't just bend them in any direction... it was a really nice geometric problem. The tunnel is a double curvature, and you have these point-clouds ... there were loads of quite nice little problems and you can get lost in it.

PB: It doesn't sound like a boring project?

PL: No it's absolutely not boring, it's just funny. None of us have worked in rail before. No one has ever worked in these contexts. We just turned up and went: "Why'd you do it like that?"

Once you finally get your mesh of the tunnel, what you're trying to do is subdivide that mesh into geometry again, another nice problem. A grid subdivision or triangles or hexagons, my personal favorite. And then you're trying to work out, which one of these grid subdivisions contains already a signal box, a cable or another obstruction basically? What sort of degree of freedom do I have to navigate through this? Taking a very detailed sub-millimeter accuracy point-cloud, that you're reducing into a subdivision of squares, simplifying it right down. And then you turn it into an evaluation. And then you have a path-finding algorithm that tries to join all the bits together within the engineering rules, within how you bend the cable. And you can imagine that by the time you get to start mesh subdivision, if you process that input to death, it's going to be absolutely meaningless. It will work in the sense that it will run and it will look quite cool, but what do I do with it?

I try to talk about language with everybody

PL: I try to talk a little bit about language with everybody. I'm trying not to overburden everybody with all of my predilections. I can't really impose anything on them. Language is a big thing, like explaining Genetic Algorithms with phrases like , "So this is the population that would kill everybody, that's like unsuitable or invalid." For example if you use a multi objective Genetic Algorithm, you might try to keep an entire set of all solutions or *configurations*, as we would call them, that you create through all of the generations of the process. The scientific language for this is "population". That's how you have to talk. You might say, "I have a population of fifty generations of the algorithm. Five thousand individuals would be created throughout the whole process. And in each generation you're only "breeding" or combining a certain set and you discard the others." You leave them behind, that's quite common. And we had a long talk about whether or not we should keep all of the things that were created and the discussion was going on like, "But some of them were just like rubbish. They're just stupid. We should just kill them, no one needs to see them again." And I'm like, "well I don't know, I quite like to keep everybody!"

Of course all you're really doing is optimizing, tending towards something that's better and you lose the possibility of chance and miss something. There's a massive bit of randomness in it and you have a whole set of controls about how much randomness you allow through the generational processes and so I have this massive metaphor and it comes with huge, problematic language around genetics and all that kind of stuff that is encoded with even more problematic language, without any criticality, into the algorithmic process. And then someone is telling me "I've got a slider that says increase the randomness on that". So it's full of all those things, which I find very challenging.

But if you ever could strip away from the language and all of the kinds of problems, you look at it in purely just what it does, it's still interesting as a process and it can be useful, but the problem is not what the algorithm does. It's what culturally those algorithms have come to represent in people's imagination.

The Hairy Hominid effect

PB: We would like to bring up the Truthful Hairy Hominid here.[4] The figure emerged when looking over the shoulder of a designer using a combination of modeling softwares to update the representation of human species for the "Gallery of Humankind". They were working on one concrete specimen and the designer was modeling their hair, that was then going to be placed on the skin. And someone in our group asked the designer, "How do you know when to stop? How many hairs do you put on that face, on that body?" And then the designer explained that there's a scientific committee of the museum that handed him some books, that had some information that was scientifically verified, but that all the rest was basically an invention. So he said that it's more or less this amount of hair or this color, this density of hair. And this is what we kept with us: When this representation is finished, when the model is done and brought from the basement to the gallery of the museum, that representation becomes the evidence of truth, of scientific truth.[5]

PL: It acts like as a stabilization of all of those thoughts, scientific or not, and by making it in that way, it formalizes them and becomes unchallengeable.

PB: The sudden arrival of an invented representation of hominids on the floor of a natural science museum, this functional invention, this efficacy, is turned into scientific truth. This is what we call "The Hairy Hominid effect". Maybe you have some stories related to this effect, on the intervention of what counts or what is accountable, what counts? Or is the tunnel already one?

PL: Well, the technology of the tunnel project maybe is, and how we're using this stuff.

A point-cloud contains millions and millions of data-points from surveys, like in LiDAR scanning, it's still really novel to use them in our industry, even though the technology has been around for years.[6] I would say the reason it still gets pushed as a thing is because it has become a massive market for proprietary software companies who say: "Hey, look, you have this really cool map, this really cool point cloud, wouldn't it be cool if you could actually use it in a meaningful way?" And everybody goes, "yes!", because these files are each four and a half gigabytes, you need a three thousand pound laptop to open it and it's not even a mesh, you can't really do anything with it, it just looks kind of cool. So the software companies go: "Don't worry about it. We'll sell you thousands of pounds worth of software, which will process this for you into a mesh". But no one really is thinking about, well... how do you really process that?

A point-cloud is just as a collection of random points. You can understand that it is a tunnel or a school or a church by looking at it, but when you try and get in there and measure, if you're really trying to measure a point cloud ... what point do you choose to measure? And whilst they say the precision is like plus or minus zero point five millimeters... well, if that was true, why have we got so much noise?

The only thing that's real are the data points

PL: One of the things that everybody that everybody thinks is useful, is to do object classification on a point-cloud, to find out what's a pipe, what's a light, what's a desk, what's a chair. To isolate only those points that you see and then put them on a separate layer in the model and isolate all those things by category. The way that that's mostly done right now, even in expensive proprietary software, is manually. So somebody sits there and puts a digital lasso around a bunch of points. But then how many of the points, when did you stop, how did you choose how to stop? Imagine, processing ten kilometer of tunnel manually...

PB: It's nicer to go around with spray paint then.

PL: Definitely.

The most extensive object classification techniques come from autonomous vehicles now, that's the biggest thing. These data-sets are very commonly shared and they do enough to say, "This is probably a car or this is a sign that probably says this" but everything is guesswork. Just because a computer can just about recognize some things, it is not vision. I always think that computer vision is the wrong term. They should have called it computer perception.

There is a conflation between the uses of computer perception for object classification, around what even *is* an object and anyway, who really cares whether it's this type or this, what's it all for?

Conflating object classification with point-cloud technology as a sup-posedly perfect representation, is actually useless because you can't identify the objects that you need and anyway, it has all these gaps, because it can't see through things and then there is a series of meth-ods, to turn that into truth, you de-noise by only sampling one in every three points... You do all of these things to turn it into something that is 'true'. That's really what it is like. It's a conflation of what's real while the only thing that's real are the data points, because well, it did capture those points.

A potential for possibilities

PB: When we spoke in Toronto six years ago, you defended generative procedures against parametric approaches, which disguise the *proba-ble* as *possible*. Did something change in your relation to the genera-tive and it's potentially transformative potential?

PL: I think it became more complex for me, when you actually have to do it in real life. I still think that there's huge risks in both approaches and at the time I probably thought that the reward is not worth the risk in parametric approaches. If you can be good at the generative thing, that's riskier, it's much easier to be bad at it, but the potential for possibilities is much higher.

What is more clear now is that these are general processes that you have to encounter, because everybody else is doing it, the bad ones are doing it. And I think it's a territory that I'm not prepared to give up, that I don't want to encounter these topics on their terms. I don't consider the manifestations, those that we don't like in lots of different ways, to be the only way to use this technology. I don't con-sider them to be the intellectual owners of it either. I am not prepared to walk away from these techniques. I want to challenge what they are in some way.

Over the last few years of building things for people, and working with clients, and having to build while we were also trying to build a group of people to work together, you realize that the parametric or procedural approaches give you an opportunity to focus on what is necessary, to clarify the decision-making in all these choices you make. It is more useful in that sense. I was probably quite surprised how little people really wanted to think about those things in genera-tive processes. So we had to start a little bit more simple.

You have to really think first of all, what is it you're going to make? Is it okay to make it? There's a lower limit almost of what's okay to make in a parametric tool because changes are really hard, because you lock in so many rules and relationships. The model can be just as complicated in a generative process, but you need to have a kind of fixed idea of what the relationships represent within your model, within your process. Whereas in generative processes, because of the very nature of the levels of abstraction, which cause problems, there are also opportunities. So without changing the code, you can just say, well, this thing actually is talking about something completely differ-ent. If you understand the math of it, you can assign a different name to that variable in your own mind, right? You don't even need to change the code.

With a parametric approach you're never going to get out of the fact that it is about a building of a certain type, you can never escape that. And we built parametric tools to design housing schemes or schools as well as some other infrastructure things, data centers even, and that is kind of okay, because the rules are not controversial when you think about schools for example. And you're probably thinking, hang on a minute, Phil, these can be controversial, but in the context of our problem definition, they were unchallengeable by any-body, they came from the government.

Showing the real consequences

PL: Parametric approaches make problems in the rules and processes visible. I think that's a huge thing. Because of the kind of projects we are building, we are given a very hard set of rules that no one is allowed to challenge. So you try and encode them into a parametric system and it won't work basically.

In the transport infrastructure projects we were doing, there are rule changes with safety, like distance between certain things. And we could show what the real consequences would be. And that this was not going to achieve the kind of safety outcome that they were looking for. Sometimes you're just making it very clear what it was that they thought that they were asking for. You told us to do it like this, this is what it gives you, I don't think that's what you intended.

We never allowed the computer to solve those problems in any of the things we've built. It just tells you, just so you know, that did not work, that option. And that's very controversial, people often don't really like that. They're always asking us to constrain them. "Why do you let the system make a mistake?"

Sometimes it is not better than nothing

PB: When we speak to people that work with volumetric systems, whether on the level of large scale databases for plants, or for making biomedical systems ... when we push back on their assumption that this is reality, they will say, "Of course the point-cloud is not a reality. Of course the algorithm cannot represent population or desire." But then when the system needs to work, it is apparently easy to let go of what that means. The need to make it work, erases the possibility for critique.

PL: One of the common responses I see is something like, "Yeah, but it is better than nothing." Or that is at least part of the story. They have a very Modernist idea that you run this linear trajectory towards complete know-how or knowledge or whatever and that these systems are incomplete rather than imperfect and that if you have a bit more time, you'll get there. But where we are now, it's still better than then. So why not use it?

In the construction sector you constantly encounter these unlucky wanna-be Silicon Valley tech billionaires, who will just say like, "But you just do it with a computer. Just do it with an algorithm!" They've fallen for that capitalist idea that technology will always work in the end. It *must* work. And whenever I present my work in conferences, I always talk about my team, what people are in the team, how we built it in some way. To the point that actually lot's of people get bored of it. Other people when they talk about these kinds of techniques will say "We've got this bright kid he's got a PhD from wherever. He's brilliant. He just sits in the corner. He's just brilliant." And of course, it's always a guy as well. They instrumentalize these people, as the device to execute their dream, which is that the computer will do everything. There's still this kind of a massively Modernist idea that it's just a matter of time until we get to that.

Sometimes a point-cloud is *not* better than nothing because it gives you a whole other problem to deal with, another idea of reality to process. And by the time you get into something that's usable, it has tricked you into thinking that it's real. And that's true about the algorithms as well. You're wrestling with very complicated processes and by the time you think that you kind of control it, it just controlled you, it made you change your idea of the problem. You simplify your own problem in order that you can have a process act on it. And if you're not conscious about how you're simplifying your problem in order to allow these things to act on it, if you're not transparent about that, if you don't acknowledge it, then you have a very difficult relationship with your work.

Supposed scientific reality

PL: We use genetic algorithms on a couple of projects now and the client in one project was just not interested in what methods we were using. They did not want us to tell them, they did not care. They wanted us to show what it does and then talk about that, which is kind of okay. It's anyway, not their job. The second client was absolutely not like that at all, they were looking for a full explanation of everything that we did. And our explanation did not satisfy them because it didn't fit with their dream of what a genetic process does.

We were fighting this perception that as soon as you use this technique, why doesn't it work out of the box? And then we're building this thing over a matter of weeks and it was super impressive how far we got, but he still told us, I don't understand why this isn't finished. But it took the US military fifty years to make any of this. Give me a break!

PB: The tale of genetics comes with its own promise, the promise of a closed circuit. I don't know if you follow any of the critiques on genetics from microchimerism or epigenetics, basically anything that brings complexity. They ask: what are the material conditions in which that process actually takes place? It's of course never going to work perfectly.

PL: The myth-making comes with the weight of all other kinds of science and therefore implies that this thing should work. Neural networks have this as well, because of, again, this storytelling about the science of it and I think the challenge for those generative processes is exactly in their link to supposed scientific realities and the sort of one-to-one mapping between incomplete science, or unsatisfactory science, into another incomplete unsatisfactorily discipline, without question. You can end up in pretty spooky place with something like a Genetic Algorithm that is abstracted from biochemistry, arguing in a sort of eugenic way.

You can only build one building

PL: I think inherent in all of this science is the idea that there is a right answer, a singular right answer. I think that's what optimization means. For the sort of stuff we build, we never say, "This is the best way of doing it." The last mile of the process has to be a human that either finishes it, fills in the gaps or chooses from the selection that is provided. We never ever give one answer.

I think someone in my world would say, "But Phil, we're trying to build a building, so obviously we can only build one of them?" This is not quite what I mean, I think there's an idea within all of the scientific constructs of the second half of the twentieth century where computer vision and perception, computer intelligence, whatever you want to call it, and genetics, they're the two biggest things. Within both of those fields, there's the idea that we will know, that we will at some point find out a truth about ourselves as humans and about ourselves plus machines. And we will make machines that look like us and then tell ourselves that because the machine performs like this, we are like those machines. I think it's a tendency which is just super Modernist.

They want a laser line to get to the best answer, the right answer. But in order to get to that, the thing that troubles me probably most of all, and this is true in all of these systems whether parametric or genetic, is the way in which the system assumes a degree of homogeneity.

It does not really matter that it is ultimately constrained

PL: I think with these generative algorithmic processes, people don't accept constraint either discursively or even scientifically. At most they would talk about the moment of constraint being beyond the horizon of usefulness. At some point, it doesn't create every possible combination. Lots of people think that it can create every option that you could ever think of. Other people would say that it is not infinite, but it goes beyond the boundary of what you would call, 'the useful extent of your solution space', which is the kind of terminology they use. I think that there's a myth that exists, that through a generative process, you can have whatever you want. And I have been in meetings where we showed clients something that we've done and they say, "Oh, so you just generated all possible options." But that's not quite what we did last week!

There's still that sort of myth-making around genetic algorithms, there's an illusion there. And I think there's a refusal to acknowledge that the boundary of that solution space is set not really by the process of generation. It's set at the beginning, by the way in which you define the stuff that you act on, through your algorithmic process.

I think that's true of parametrics as well, it's just that it's more obviously to improve metrics. Like, here's a thing that affects this thing. And whether you complexify the relationships between those parameters, it doesn't really matter, it's still kind of conceptually very easy to understand. No matter how complex you make the relations between those parameters, you can still get your head around it. Whereas the generative process is a black box to a certain extent, no one really knows, and the constraint is always going to be on the horizon of useful possibilities. So it doesn't really matter that it is ultimately constrained.

We're not behaving like trained software developers

PL: By now we have about twenty people on our team and they're almost all architects.

When I do a presentation in a professional context, I have a slide that says, "We're not software developers, but we do make software." And then I try to talk about how the fact that we're not trained as software developers, means that we think about things in different ways. We don't behave like them. We don't have these normative behaviors from software engineering in terms of either what we create or in the way in which we create things. And as we grow, we make more things that you could describe as software, rather than toolkits or workflows.

After one of these events, someone came up to me and said, "Thank you, that was a very interesting talk. And then she asked, "So who does your software development? To whom do you outsource the development?" It was completely alien to this person that our industry could be responsible for the creation of software itself. We are merely the recipients of product satisfaction.

Architects are not learning enough about computation technology either practically or critically, because we've been kind of infantilized to be the recipient discipline.

Not everyone can take part

PB: We noticed a redistribution of responsibilities and also a redistribution of subjectivity, which seems to be reduced to a binary choice between either developer or user.

PL: I think that's true. It feels like we're back in the early nineties actually. When computational technology emerged into everyday life, it was largely unknown or was unknowable at the time to the receiving audience, to the consumers. There was a separation, an us and them, and even talking about a terrible version of Windows or Word or something, people were still understanding it as something that came from this place over there. Over the last two or three decades, those two things are brought together, and it feels much more horizontal; anybody can be a programmer. And now we're back at the place where we realize that not everybody can be part of creating these things at all.

Governments have this idea that we'll all be programmers at some point. But no, we won't, that's absolutely not true! Not everybody's going to learn. So one of the things I try to specifically hold on is that we need to bring computational technology to our industry, rather than have it created by somebody else and then imposed on us.

The goal is not to learn how to be all a software company or a tech company.

If something will work, why not use it?

PB: We are troubled by the way 3D techniques and technologies travel from one discipline to another. It feels almost impossible to stop and ask "hey, what decisions are being made here?" So we wanted to ask you about your experience with the intense circulation of knowledge, techniques, devices and tools in volumetric practice.

PL: It is something that I see every day, in our industry, and in our practice. We have quite a few arguments about the use of image recognition or facial recognition technologies for example.

When technologies translate into another discipline, into another job almost, you don't just lose the ability to critique it, but it actually enhances its status by that move. When you reuse some existing technology, people think you must be so clever to re-apply it in a new context. In the UK there are tons of examples of R&D government funding that would encourage you to use established existing techniques and technologies from other sectors and reapply them in design and construction. They don't want you to reinvent something and they certainly don't want you to challenge anything. You're part of the process of validating it and you're validated by it. And similarly, the people of the originating discipline get to say, "Look how widely used the thing we created is", and then it becomes a reinforcement of those disciplines. I think that it's a huge problem for anyone's ability to build critical practices towards these technologies.

That moment of transition from one field to another creates the magic, right? A technology apparently appears out of nowhere, lands fully formed almost without friction and without history. It lacks history in the academic sense, the scientific process and indeed, also lacks the labor of all of the bodies, the people that it took to make it, no one cares anymore at that point.

What I've seen in the last five years is that proprietary software companies are pushing things like face recognition and object classification into Graphical User Interfaces (GUIs), into desktop software. Something like a GAN or whatever is not a button and not a product; it is a TensorFlow routine or a bunch of Python scripts that you get off GitHub.

There's a myth-making around this, that makes you feel like you're still engaged in the kind of practice of creating the technique. But you're not, you're just consuming it. It's ready-made there for you. Because it sits on GitHub, you feel like a real coder, right? I think the recipient context becomes infantilized because you're not encouraged to actually create it yourself.

You're presented with something that will work, so why not use it? But this means you also consume all of their thinking all of their ways of looking at the world.

Notes

1. ↑ See: Possible Bodies, "MakeHuman," in this book.
2. ↑ See: "Phil Langly in conversation with Possible Bodies, Comprehensive Features," https://volumetricregimes.xyz/index.php?title=Comprehensive_Features.
3. ↑ TensorFlow is "An end-to-end open source machine learning platform" used for both research and production at Google. https://www.tensorflow.org/.
4. ↑ "Item 086: The Truthful Hairy Hominid," *The Possible Bodies Inventory*, 2014.
5. ↑ Another aspect of the *Hairy Hominid effect* appears in our conversation with Simone C Niquille, "The Fragility of Life," in this chapter.
6. ↑ LiDAR is an acronym of "light detection and ranging" or "laser imaging, detection, and ranging".

Somatopologies: On the ongoing rendering of corpo-realities

Clumsy Volumetrics

Helen V. Pritchard

Opening the on-line *Possible Bodies Inventory*, we encounter an abundance of items — shifting numbered entries of manuals, mathematical concepts, art-projects and micro-CT images of volumetric presences. So-called bodies in the context of hardware for scanning, tracking, capturing and of software tools for data processing and 3D-visualization. Working on-and-with the *Possible Bodies Inventory* is an inquiry on the materialization of bodies and spaces, in dizzying relation with volume practices. As discussed throughout this book, the volumetric regime directs what so-called bodies are — and how they are "shaped by the lines they follow".[1] As Sara Ahmed outlines in her queer phenomenology, orientations matter in how they shape what becomes socially as well as bodily given; that is how bodies materialize and take shape.[2] Many items in the *Possible Bodies Inventory* evidence how the orientations of 3D practices matter significantly in materializing spaces for bodies that are inhabitable for some, and not others.[3] Rocha and Snelting refer to this as the the "very probable colonial, capitalist, hetero-patriarchal, ableist and positivist topology of contemporary volumetrics."[4] Indeed the *Possible Bodies Inventory* demonstrates how the inherited histories of colonialism stretch into 3D practices to shape and direct bodies: "colonialism makes the world 'white', which is of course a world 'ready' for certain kinds of bodies, as a world that puts certain objects within their reach".[5] This orientation starts within the worldsetting of x = 0, y = 0, z = 0 and spreads out across 3D space; the mesh, the coordinate system, geometry and finally, the world.[6] However, what are the orientations that spread from this computational world-setting to shape spaces? How does it also reinforce what is already made reachable or not, livable or not, from what Louis Althusser calls the zero point of orientation, from which the world unfolds?[7] As Possible Bodies observe in *Item 007: Worldsettings for beginners*:

> *Using software manuals as probes into computational*
> *realities, we traced the concept of "world" in Blender, a*
> *powerful Free, Libre and Open Source 3D creation suite.*
> *We tried to experience its process of "worlding" by staying*
> *on the cusp of "entering" into the software. Keeping a*

*balance between comprehension and confusion, we used
the sense of dis-orientation that shifting understandings
of the word "world" created, to gauge what happens when
such a heady term is lifted from colloquial language to be
re-normalized and re-naturalized. If the point of origin
changes, the world moves but the body doesn't.*

As Possible Bodies feel-out, in their software critique of 3D graphics
software *Blender*, in volumetric regimes, when worlds are set, the
possibilities for bodies are narrowly scripted — computationally pre-
determining the objects that stay in reach. And like in the physical
world these "orientations become socially given by being repeated
over time".[8] Indeed, as *Item 007* shows, volumetric world-settings are
an attempt to fix in place how the world unfolds from a zero-point orientation. An orientation which shapes and is shaped by a certain kind
of body as a norm and what Ahmed calls less room to wiggle — "[l]ess
wiggle room: less freedom to be; less being to free".[9] So, in volumetric
regimes — when worlds are *world-set* in ways that computationally
shape the body to the world, through directions between fixed points,
what about the bodies that don't fit or don't follow the set directions?

Ahmed suggests that "clumsiness" might be the way to form a
queer and crip ethics to generate new openings and possibilities.
Clumsy referring to when we wiggle off the path, are out of time with
each other and become in the way of ourselves:

*Bodies that wriggle might be crip bodies, as well as a
queer bodies; bodies that do not straighten themselves
out. The elimination of wriggle might be one form of what
Robert McRuer calls "compulsory able-bodied-ness,"[10]
which is tied to compulsory [cis-gendered] straightness,
to being able to follow as closely as you can the line you
are supposed to follow.[11]*

Making the affinity present between queer and crip, Ahmed notes,
clumsiness is not always a process which brings us together or
attunes us, it can also be the moments – the desiring moments – when
we bump into the world. Clumsiness is a powerful political orientation, one in which our ways of relating to, and depending on, each
other are reconfigured, promising as McRuer notes, possibilities to
"somehow access other worlds and futures".[12] By awkwardly reaching
towards some of the items at the inventory, can we orient volumetric
practices that make wiggle room, deviate from straightness and open
up new liberatory paths? Informed by the difficulties of following the

paths of queer life and world-declarations, might we form paths of queer desire for bodies? Such desire might pass through tentative processes to de-universalize, de-centralize, de-compose and re-visit tools and practices in order to better understand the conditions of their mutual constitution. Paths made through workarounds, interventions and hacks of volumetric hardwares and softwares that deviate from social-givens.

Queerness matters because it affects what we can do, where we can go, how we are perceived, and so on. Yet we also know about creative wiggles, wiggling off paths when our bodies don't fit and the queer wiggle of wiggling in cramped spaces.[13] Ahmed writes that for queers "it is hard to simply stay on course because love is also what gives us a certain direction" creating orientations of desire that generate new shapes and new impressions.[14] However, although love might give us a certain direction, it can take a lot of work to switch orientations. Turning towards a queer ethics of clumsiness for volumetrics then might take some work to make room for non-attunement, not seeing this as a loss of possibility but as opening new paths; making accounts of the damages done to bodies who stray from the world-settings of volumetric regimes; and unfolding new ways which bodies shape and are shaped by calculations.

As the disobedient action research of *Item 007* demonstrates, in computer graphics and other geometry-related data processing, calculations are based on Cartesian coordinates, consisting of three different dimensional axis: x, y and z. In 3D-modelling, this is what is referred to as "the world".[15] The point of origin literally figures as the beginning of the local or global computational context that a 3D object functions. But what is this world that is set and how does it shape or is shaped by so-called bodies? In a discussion of facial reconstruction by forensic science, Vicki Kirby, drawing on Bruno Latour's work on scientific reference, suggests we would be wrong to assume that the relationships conjured in 3D modelling are simply an illusion or mirror.[16] Instead, Kirby demonstrates that there is a relationship between 3D models and the physical world, what she calls communicative intimacies and peculiar correspondences, that are conjured between a 3D modelled face and the data gathered from a fragment of a skull.[17] That is to say there is often some resonance between data collected in one site and modelled or visualized in another, which opens up the possibility for agency in 3D. Forensic science practices are based on techniques that pre-date computers, but that are refined by the use of ultrasound data from living people, computed tomography (CT scans), and magnetic resonance imaging (MRIs) and Kirby shows how:

> *data taken from one temporal and spatial location can contain information about another; a fragment of skull is also a sign of the whole, just as an individual skull seems to be a specific expression of a universal faciality. In other words, there is no simple presence versus absence in these examples.*[18]

Kirby proposes (following Latour) that this is because the world, as a more-than-human assembly, has the capacity to produce nodes of reference, or evidence, that effectively correspond.[19] That is, *the world is present* in 3D scans and models. Kirby suggests this means we need then to consider the possibility of the peculiar correspondence between the physical world and 3D models not as loss, or reduction of nature/world but as its playful affirmation.[20] This recognition *does* open up the powerful possibility for a 3D practice which is understood as inhabited by the liveliness of the world. However what Kirby *does not* acknowledge is that because the world *is* present in 3D practices, they are also already materially oriented towards social givens of what faces (or forests) are. This is particularly poignant in the model of an "evolutionary" body type facial reconstruction documented in *Item 086: The Truthful Hairy Hominid.* The item shows us documentation from an excursion to the basement of the Natural Sciences Museum in Brussels, highlighting the dependence of 3D practices of facial reconstruction on scientific racism. This is also evidenced in the research of Abigail Nieves Delgado, who through a series of semi-structured interviews with experts in facial reconstruction, shows how "when reconstructing a face, experts carry out a particular way of seeing [...] that interprets visible differences in bodies as racial differences".[21] She suggests that this analysis highlights that facial reconstructions should be understood as objects that allow us to trace past and present pathways of racial thinking in science. Delgado shows how the scientists and modellers she interviewed see skull shapes as part of specific narratives about purity, mixture, nation and race, narratives that reiterate the violence of scientific racism. Delgado argues that "by looking at facial reconstruction, we also learn that to stop reproducing race means to stop seeing [and modelling] in racial terms"[22] — a way of seeing based on racialized categories that have become embedded within scientific practices as neutral. This

normative seeing is held in place by 3D volumetrics and facial reconstruction practice. So, whilst we might recognize that the world is
present in 3D models and this opens up possibilities for encountering
the liveliness of the world, we also need to recognize these same models are informed by the inherited histories of the sciences in which
they operate.

Alongside the violent directive softwares and hardwares of the
industrial continuum of volumetric regimes, the *Possible Bodies
Inventory* also holds and sorts propositions that hold the liveliness of
the world, its shaping capacities and find ways to remake volumetrics,
destabilizing the inherited histories of colonialism, ableism and
racism within the sciences that inform 3D practices. The queer and
crip volumes are full of the pleasure, tenderness and excitement of
opening worlds. Hacked scanners, misused models, lumpy bodies all
create glimmering deviations, which rotate as alternative volumetrics. These inventory items generate the proposal of working with
other references within 3D modelling, held in tension with the technical aspects of 3D modelling. Or as Snelting discusses:

> *we might use awkwardness to move beyond thinking
> about software in terms of control using awkwardness as
> a strategy to cause interference, to create pivotal
> moments between falling and moving, an awkward in-
> between that makes space for thinking without stopping
> us to act.*[23]

This pleasurable, loving, reorientating between falling and moving in
the *Possible Bodies Inventory* includes inventory items that make
present volumes generated by human and more-than-human bodies
such as scanner, flowers, plants, trees, human gestures, minerals and
anatomy. Working on and with these inventory items is alike to what
Jas Rault and T. L. Cowan describe as entering into a collective deep
queer processing — "the possibilities for understanding process as a
sexy, sometimes agonized but always committed, method, an orientation towards unruly information".[24]

One of these orientations towards unruly information is *Item
035: Difficult Forests* by Sina Seifee, *Difficult Forests* turns us to moving coordinates, colors and what Seifee describes as *Memoirs*.[25] In
2013 Seifee travelled to the Amazon region in Colombia with the
Kinect as a recording device. The Kinect was hacked to work as a kind
of LiDAR to create a series of digital memoirs spelled out as systematic screen glitches, technological relationships and life histories:

165

The representation of the journey — itself as complex
problematic event — together with the horde of visual
artifacts tell a set of interfacial stories with my co-trav-
ellers. This project addresses the splicing of direct and
tactile human perception of reality with another reality,
one that is mediated and technical. It is an aesthetic
dream, dream of isomorphism between the discursive
object and the visible object in the Amazonian forests.[26]

Difficult Forests generates queer traces of desire, the images and text
creating different routes to get to this point or to that point. Here
deviating in the forest resets stability and make new co-ordinates of
points between so-called bodies — they wiggle from the 0,0,0 of
worldsetting. Seifee discusses how sometimes the Kinect is held by
him, sometimes by his companion or the 3 year old with them. Destabilizing the imaginary of the lone able-bodied cis male scientist who
scans the forest under difficult conditions, the different paths
become queer intergenerational "multiple world-declarations".[27]
Using the hacked Kinect to generate measurements from a zero point
that is never still, *Item 035* opens up the possibility for the movement
between points to be queered, to be reinhabited and change course,
whilst not letting go of the possibilities of volumetric knowledge production. The Kinect extends the reach of the body, whose bodies reach
and the forest. Seifee documents this extension of reach in the images
and text, recording how the body becomes-with the difficult forest as
it takes in that which is "not" it. What Ahmed describes as the "the
acquisition of new capacities and directions — becoming, in other
words, "not" simply what I am "not" but what I can "have" and "do". The
"not me" is incorporated into the body, extending its reach".[28] These
more-than-human capacities and directions shape forest, scanner
and body. As Seifee notes, "The forest recorded and screen captured
while walking in a "directly lived" space — in sweat, heat, fatigue and
mosquito bite".[29] The result is a corrupted Kinect scan of the forest,
where the mapped surfaces of leaves float around a body without stable ground, as the forest unfolds. It asks us to consider the practice of
3D scanning as a practice of memoir in which the world is made
present as a shaping that unfolds through surface encounters (rather
than linear methods of collection).

In these memoirs *Difficult Forests* seems marked with details
that are the "indelible and complex entanglements of nature/cul-
ture".[30] The memoirs of *Difficult Forests*, are more-than-human and

dazzling with the reticulant agency of the forest. As Seifee notes, "the Amazon rainforest still resists to remain a radical nonhuman surrounding on the surface of the earth".[31] The memoirs problematize the overlapping surface and jungle, yet the result is not a visual without reference — both scientific and affective. The images correspond to a set of measured points and the forest is still present in a felt shaping way. We witness the dense and lively agency of the forest and the human-machine scanners in this unstable scan. *Difficult Forests* reminds us that there is a possibility to conjure a looser translation between local and global coordinates one that stays with the openings 3D offers but also proposes new ways of seeing with 3D. It reorients the translation between the local and global (data) that emerges from 3D scanning in inventive ways — making room for deviations from set paths between points and bodies that emerge as different shapes.

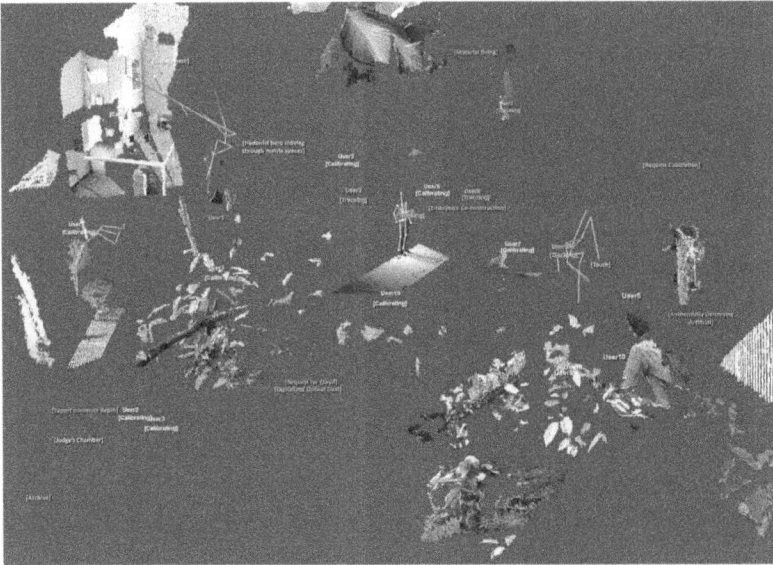

Sina Seifee, *Difficult Forests*, 2013

Scanning differently is also explored in *Item 33*, Pascale Barret's work *This obscure side of sweetness is waiting to blossom.*[32] *Item 33* is a flowering bush made present as a 3D printed object through unconventional uses of scanning devices, point clouds and surface meshes. If we tenderly hold *Item 33* in our hands, we can feel out the unfinished 3D printed edges and uncontainable volumes. The awkward lumpy mass of scanned leaves and 3D printing support structures enacts a clumsy wiggling from what has become an accepted path of 3D practices — in which objects are often presented as smoothed-off naturalized accounts or miniaturizations. Whilst still drawing lines between points, this inventory item proposes to us the possibilities for working with practices in ways that inhabit space-time of bodies-plants-scanners in a much different way. In contrast to the practice of 3D modelling which aims to capture data to recreate or reflect fixed bodies in fixed "nature", such as the 1:1 copy of a flower or leaf, this work allows an orientation in a world that is in excess of the scanner and is not made of straight lines or entities with hard boundaries. Rather than using the scanner as apparatus of colonial capture *Item 33* advocates for what Jessica Lehman calls the need to recognize volume beyond volumetrics. As Lehman outlines, "[v]olumes are irreducible to and in excess of the apparatuses of their capture, whether big science or state power".[33] A materiality that is more-than just resistant to or compliant with volumetrics. Indeed, the amalgamated movements of scanner, bodies and the plants that are shown within the 3D print make explicit the more-than-human and reticulant materiality of volume, a volume which does make present the world but is also in excess of scientific reference. An orientation towards other ways of understanding the materialization of data, practice, movement, bodies, and scanning. A volumetric practice that might provide (situated, temporary) truths about lives.

Both the degenerate Kinect scans of *Difficult Forests* and the knobbly 3D print of Barret's encounter of the blossoming bush are "volumizations" of how moving towards and getting close to objects with computation is difficult yet also shapes us — difficulty shapes us. The items makes-felt what Lauren Berlant describes as the unbearability of being oriented by objects:

> The critical object is unbearable much like the object of love is: too present, distant, enigmatic, banal, sublime, alluring and aversive; too much and too little to take in, and yet, one discovers all this only after it's been taken in, however partially, always partially, and yet overwhelmingly even at the smallest points of genuine contact.[34]

Indeed, the directing capacities of many items within the inventory bring attention to the impossibility of resolving ambivalence in our knowledge practices.

The *Possible Bodies Inventory* is a proposal to consider computation as a shaping force on bodies as well as shaped by those bodies —

but importantly as this tour has shown the room for bodies to shape volumetrics may be constrained by inherited histories and social givens. These inventory items open new paths by their wiggle work orientating away from the inherited constraints, rethinking what it means to compute volumes — generating queer and crip ethics to orient practices. As inventory items *Difficult Forests* and *This obscure side of sweetness is waiting to blossom* hint, 3D practices such as scanning might make possible smallest points of genuine contact with the materiality of the world, without demanding a stabilizing resolution or normative relations. These two items propose a type of pleasurable queer processing, a clumsy computing that works against the muscular straight lines and modes of reduction for efficiency within volumetric practices. Making-possible the presence of the world without overstabilizing paths or resolving the difficulty of contact. Generating volumes that work-with rather than against the body in motion — queer wiggles that move us towards other bodies, objects and political transformations even in tight, hard to reach spaces.

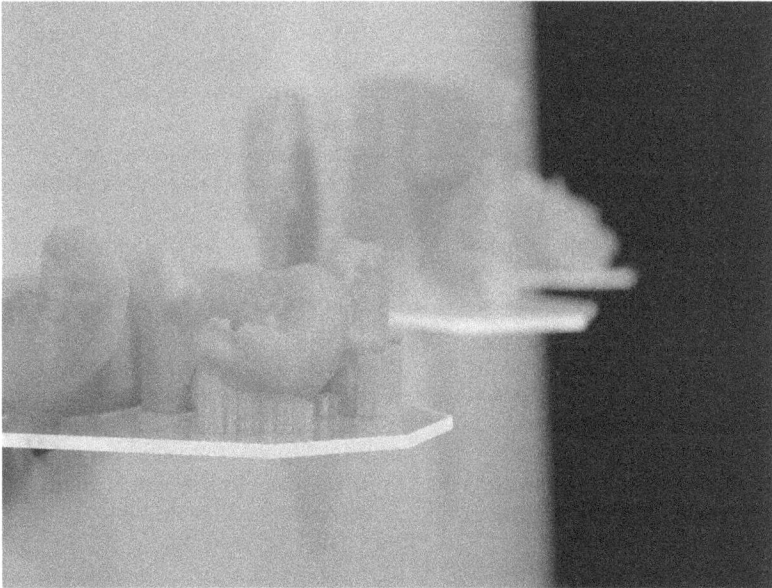

Pascale Barret, *This obscure side of sweetness is waiting to blossom*, 2017

Notes

1. ↑ Sara Ahmed, *Queer Phenomenology: Orientations, Objects, Others* (Duke University Press, 2006), 133.
2. ↑ Ahmed, *Queer Phenomenology*
3. ↑ See also for example Romi Ron Morrison, "Endured Instances of Relation, an exchange," Possible Bodies, "So-called Plants," and Jara Rocha, "Depths and Densities: A bugged report," in this book.
4. ↑ Jara Rocha, and Femke Snelting, "The Industrial Continuum of 3D," in this book.
5. ↑ Ahmed, *Queer Phenomenology*, 87.
6. ↑ Possible Bodies, "Disorientation and its Aftermath," in this book.
7. ↑ Edmund Husserl, "Ideas Pertaining to a Pure Phenomenology and to a Phenomenological Philosophy: Second Book Studies in the Phenomenology of Constitution," Vol. 3. Springer Science & Business Media, 166, cited in Ahmed, *Queer Phenomenology: Orientations, Objects, Others*, 8.
8. ↑ Ahmed, *Queer Phenomenology*, 77.
9. ↑ Sara Ahmed, "Wiggle Room," *Feministkilljoys* (blog), September 28, 2014, https://feministkilljoys.com/2014/09/28/wiggle-room/.
10. ↑ Robert McRuer, "Crip Theory: Cultural Signs of Queerness and Disability," Vol. no.9, NYU press.
11. ↑ Ahmed, "Wiggle Room".
12. ↑ McRuer, "Crip Theory," 208.
13. ↑ Ahmed, *Queer Phenomenology*, 77.
14. ↑ Ahmed, *Queer Phenomenology*, 19.
15. ↑ See: Possible Bodies, "Disorientation and its Aftermath," in this book.
16. ↑ Vicki Kirby, *Quantum Anthropologies: Life at Large* (Duke University Press, 2013), 78.
17. ↑ Kirby, *Quantum Anthropologies*, 26.
18. ↑ Kirby, *Quantum Anthropologies*, 26.
19. ↑ Kirby, *Quantum Anthropologies*, 81.
20. ↑ Kirby, *Quantum Anthropologies*, 20.
21. ↑ Abigail Nieves Delgado, "The Problematic Use of Race in Facial Reconstruction", *Science as Culture* 29, no. 4, (2020): 568.
22. ↑ Nieves Delgado, "The Problematic Use of Race in Facial Reconstruction".
23. ↑ Femke Snelting, "Awkward Gestures," in *The Mag.Net Reader 3: Processual Publishing: Actual gestures*, eds. Alessandro Ludovico and Nat Muller (London: OpenMute Press, 2008).
24. ↑ Jas Rault and T.L Cowan, "Heavy Processing for Digital Materials (More Than A Feeling): Part I: Lesbian Processing", 2020, http://www.drecollab.org/heavy-processing/[1].
25. ↑ See: Sina Seifee, "Rigging Demons," in this book.
26. ↑ Sina Seifee, "Difficult Forests," 2016, http://www.sinaseifee.com/DifficultForests.html.
27. ↑ Possible Bodies, "Disorientation and its Aftermath".
28. ↑ Ahmed, *Queer Phenomenology*, 91.
29. ↑ Sina Seifee, "Amazon Talk," FULL SATURATION in Kunstpavillon München, 2014, https://seifee.com/post/138661817448/amazon-talk.
30. ↑ Myra J. Hird, "Feminist Engagements with Matter," *Feminist Studies* 35, no. 2 (2009): 329-47.
31. ↑ Seifee, "Amazon Talk."
32. ↑ Another take on this item can be found in Jara Rocha, Femke Snelting, "So-called Plants," in this book.
33. ↑ Jessica S. Lehman, "Volumes beyond Volumetrics: A Response to Simon Dalby's "The Geopolitics of Climate Change," *Political Geography* no. 37 (2013): 51-52.

34. ↑ Lauren Berlant, "Conversation: Lauren Berlant with Dana Luciano," *Social Text Journal* online (2013), https://socialtextjournal.org/periscope_article/conversation-lauren-berlant-with-dana-luciano.

Somatopologies (materials for a movie in the making)

Possible Bodies (Jara Rocha, Femke Snelting)

Somatopologies consists of texts and 3D-renderings with diverse densities, selected from the Possible Bodies Inventory. Each of them wonders from a different perspective about the regimes of truth that converge in volumetric biomedical images. The materials investigate the coalition at work between tomography and topology that aligns math, flesh, computation, bone, anatomic science, tissue and language. When life is made all too probable, what other "bodies" can be imagined? In six sequences, Somatopologies moves through the political fictions of somatic matter. Rolling from outside to inside, from a mediated exteriority to a computed interiority and back, it reconsiders the potential of unsupervised somatic depths and (un-)invaded interiors. Unfolding along situated surfaces, this post-cinematic experiment jumps over the probable outcomes of contemporary informatics, towards the possible otherness of a mundane (after)math. It is a trans*feminist exercise in and of disobedient action-research. It cuts agential slices through technocratic paradigms in order to create hyperbolic incisions that stretch, rotate and bend Euclidean nightmares and Cartesian anxieties.

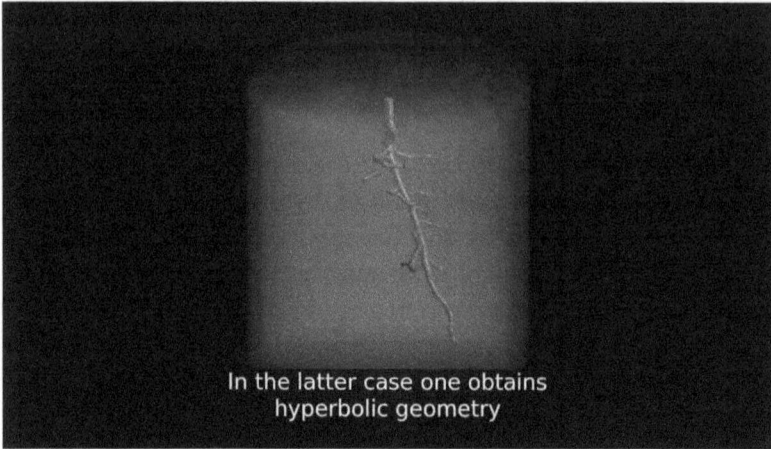

In the latter case one obtains
hyperbolic geometry

Item 005: Hyperbolic Spaces + Item 082: Ultrasonic Dreams

Non-euclidean geometry is what happens when any of the 5 axioms do not apply. It arises when either the metric requirement is relaxed, or the parallel postulate is replaced with an alternative one. In the latter case one obtains hyperbolic geometry and elliptic geometry, the traditional non-Euclidean geometries. When the metric requirement is relaxed, then there are affine planes associated with the planar algebras which give rise to kinematic geometries that have also been called non-Euclidean geometry.[1]

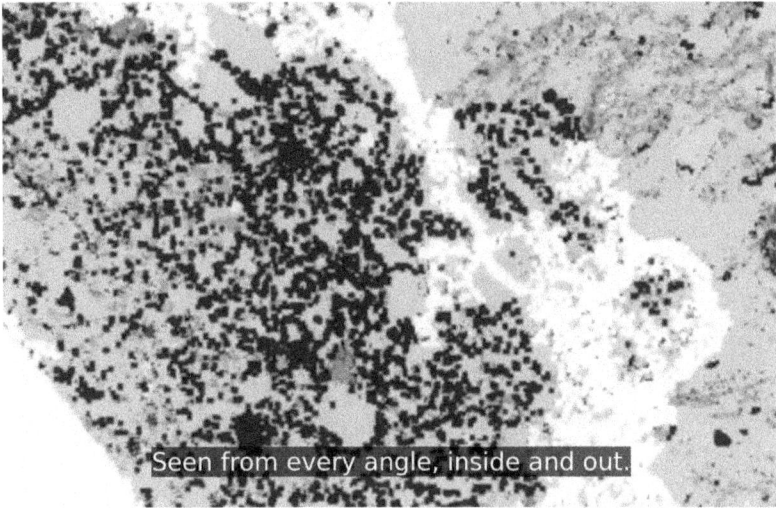

Seen from every angle, inside and out.

Item 099: Porous micro-structures + Item 071: Visible Woman

No one knows her name. Or why she ended up here. On the internet. In classrooms. In laboratories. Cut into thousands of slices. Picked over and probed. Every inch analyzed and inspected by strangers, around the world. She is the most autopsied woman on earth. The world's one and only Visible Woman has revealed everything for the sake of Modern science. Except ... her identity.[2]

Item 098: Region Of Interest + Item 028: Circlusion and/or circluding

A new term, one that has been missing for a long time: "circlusion". It denotes the antonym of penetration. It refers to the same physical process, but from the opposite perspective. Penetration means pushing something — a shaft or a nipple — into something else — a ring or a tube. Circlusion means pushing something — a ring or a tube — onto something else — a nipple or a shaft. The ring and the tube are rendered active. That's all there is to it.[3]

First things first,
find your Region Of Interest.

Item 006: The Right-Hand Rule + Item 098: Region Of Interest

First things first, find your Region Of Interest. (...) It is going to be available in all planes. Yours is not going to look like this, it might look like this: so that it surrounds the entire image. If that is the case, what you are going to do now, is drag in all four sides, so that you have basically isolated your Organ Of Interest. And you are going to do that for all the different planes as well, just so you know that we are going to get exactly what we are asking for.[4]

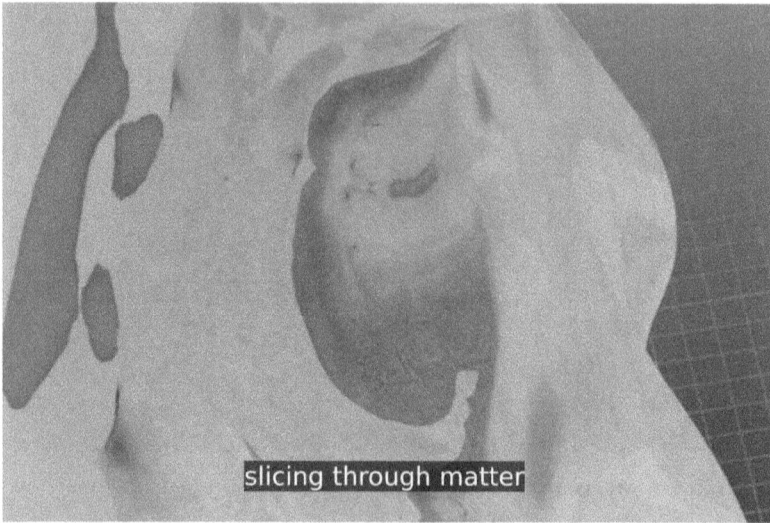

Item 017: MakeHuman + Item 082: Ultrasonic Dreams

> *Now they all moved together, more-than-human compo-*
> *nents and machines, experiencing an odd sensation of*
> *weightlessness and heaviness at the same time. Limbs*
> *stuck to the wall, atoms bristled. Bodies first lost their ori-*
> *entation and then their boundaries, melting into the fast*
> *turning tube. Radiating beams fanned out from the mid-*
> *dle, slicing through matter radically transforming it with*
> *increasing intensity as the strength of circlusion*
> *decreased.*[5]

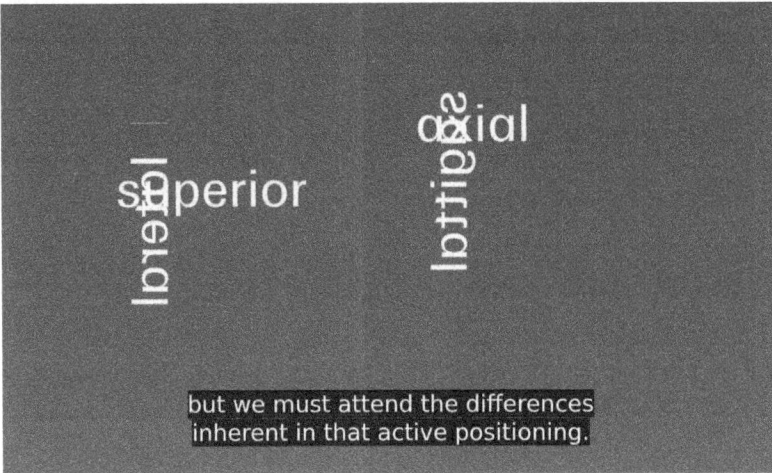

Item 070: Anatomical planes + Item 012: No Ground

Closer, further, higher, lower: the body arranges itself in perspective, but we must attend the differences inherent in that active positioning. The fact that we are dealing with an animation of a moving body implies that the dimension of time is brought into the conversation. Displacement is temporary, with a huge variation in the gradient of time from momentary to persistent.[6]

Notes

1. ↑ *Item 005* is a remix of the Wikipedia entries on: "Euclidian" and "Non-Euclidian math", inspired by the rendering of Hyperbolic Spaces in Donna J. Haraway, *Staying with the Trouble: Making Kin in the Chthulucene* (Durham: Duke University Press, 2016).

2. ↑ Transcription from "Visible Woman," American TV-documentary, 1997, https://www.youtube.com/watch?v=ZmDrlJtrByY.

3. ↑ Bini Adamczak, "On Circlusion" *Mask Magazine*, 2016, maskmagazine.com. For another take, see Kym Ward feat. Possible Bodies, "Circluding," in this book.

4. ↑ Transcription from, "Patient CT Mandible Segmentation for 3D Print Tutorial (using ITK-Snap)," 2016, https://www.youtube.com/watch?v=P44m3MZuv5A.

5. ↑ See: Possible Bodies (Helen V. Pritchard, Jara Rocha, Femke Snelting), "Ultrasonic Dreams of Aclinical Renderings," in this book.

6. ↑ See: Jara Rocha, Femke Snelting, "Dis-orientation and its Aftermath," in this book.

From Topology to Typography: A romance of 2.5D

Spec (Sophie Boiron and Pierre Huyghebaert)

This graphic contribution is based on conversations with Spec on topology and typography, which followed from their interventions in the installation *Somatopologies (materials for a movie in the making)*, Constant_V in Brussels (2018).

Soma

Nice to meet you. We are four Metapost drawings, young cousins of Metafont glyphs. Our curved bones are made of mathematical equations that our programmer thinks produce the most pleasing 2D curves, err... to his 80's eyes. The automatic closing of most shapes, intrigued Quentin Jumelin, our designer.

I'm a text block composed with the Cycle-Source font and I'm clearly not the focus point. I look less sophisticated and more common ... even if i'm one of your your children! To build me, my shapes have been expanded around the bones of my ancestor; they chose a thin stroke in two parallel borders. My corners are sharp. I'm described using Bezier curves, that were first designed to describe car parts in 3D, but twenty years later they were used for computing 2D shapes. I know that I'm just a convenience here, used to quickly compose the lay out and then be forgotten.

Soma

Is this a Z dimension, or only a Z index? The designers thought they needed our parent shapes to produce the templates. So we are back as separate shapes. We were immediately expanded following a thick width to allow the tip of a marker pass between our carved walls. It is more visible that some shapes are on top of each others. Even if there is no 3D, but a pile of 2D objects with no thickness and an order.

Soma

It would be sad if the machine that will make us templates emerge from acetate spoils energy by cutting overlapping shapes. They will fall down when the plates are moved from the grill. So some boolean operations are applied to transform our shapes mathematically, but not so much visually. Radical! Some parts are falling anyway, that's the deep nature of 3D.

Soma

Here we are, a set of words representing items from the Possible Bodies inventory. Constant's vitrine will host us. But the street bends downwards. Maybe that is why the researcher proposed to make us go up? We are happy to be composed using the font from above, hence the double lines of all our shapes that produces this weirdly thick lineweight.

Somatopologies 005. Hyperbolic spaces 006. The eyes of the rock 012. No Ground 017. MakeHuman 028. Circlusion and/or circluding 070. Anatomical planes 071. Visible Woman 082. Ultrasonic dreams of aclinical renderings 098. Region Of Interest 095. Shiny Bones 099. Porous mai

8py

wHC

AZ95087
6X

lFKir

hbtd

VGR

NTMP

eucas

AUOEI

nmwo

What to follow? We marks try to guide the letterers through a chain of operation: substraction-destruction-declutching. The designers have optimised the space on the templates, plotting only the needed letters, then they were intertwined, merged, composed. The plan is to build, but first you have to dismantle things (the modularity of things) towards more and more situated and less and less abstract shapes and countershapes.

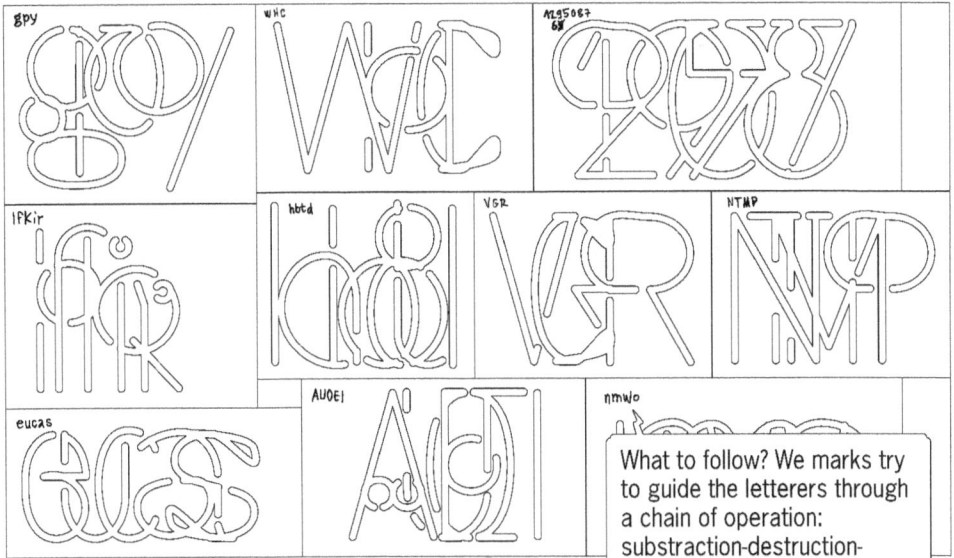

Us templates, we like to be optimized and to serve. We understand budget limitations. The overlap that the designers chose is a bit baroque to our taste but in knot-theory there are more than three axes, there's time! And there's one in front and one in the back of a rope. Curved lines, dimensions ... implicated understandings of multidimensionality. So there is no end to the rope, a ligature-writing; a flow that is there, but that is not represented. Time and sequence are more present. Speaking about time, that moment in the lasercutter was not fun and it smelled bad. Now we have been marked and let's hope that our apparel will be rigid enough to hold multiple lettering operations.

And I'm the ink deposit of the POSCA marker. I'm both mineral and pigment, my solvant has nearly evoporated. The hand of the letterer has started my journey at a certain angle at the end of the trench. I don't know if it is intentional? The pressing of the nib has left a bit more matter there. I flow languidly, guided by the linear scarred and rounded edge of the template. Operations of through, behind, inside, outside. 2.5D is a rough operation that happens in three dimensions. Combining 2D with proto-3D stuff. Borders have an on-off relationship to 3D. It's liminal, a praxis at the edge of 3D. The 2mm thickness of the material as a dimension. Can we call this the periphery of 3D? Should this be called 2.5D, even? I'm not going to settle on that question, I prefer the everlasting transition. Now I'm curious to see how the hand will do the crossing operation of an angled stroke in a few seconds, as the letter implies to continue straight to the bottom.

The hand of the person who draws or writes me constrains my 3D movement in an unprecedented way. It takes some time to navigate this new coordinate system.

Paradoxally, they forgot me when they reduced the glyphset, so they are now forced to draw me from curved shapes recycled from other letters.

Soma

This morning, I arrived with the Spec team charged with markers, and a set of 14 acrylic templates, hand-size. They started with two gosettes from De Weerdt bakery and a small coffee served by Verschueren, a bar located 20 meters from the Constant vitrine. Then, they placed me vertically against the window.I was leaning sideways on an oblique surface and forward against a freshly painted facade. For six hours, their bodies climbed me, the ladder, again and again, up and down, leaving typographic traces on a vertical plane of transparent glass. While they situatedly collapsed a stack of glyphs onto a flat surface, they kept tracing them separately and found interesting connections between them.

Here's the tool to re-use and modify!
possiblebodies.constantvzw.org/
inventory/?100

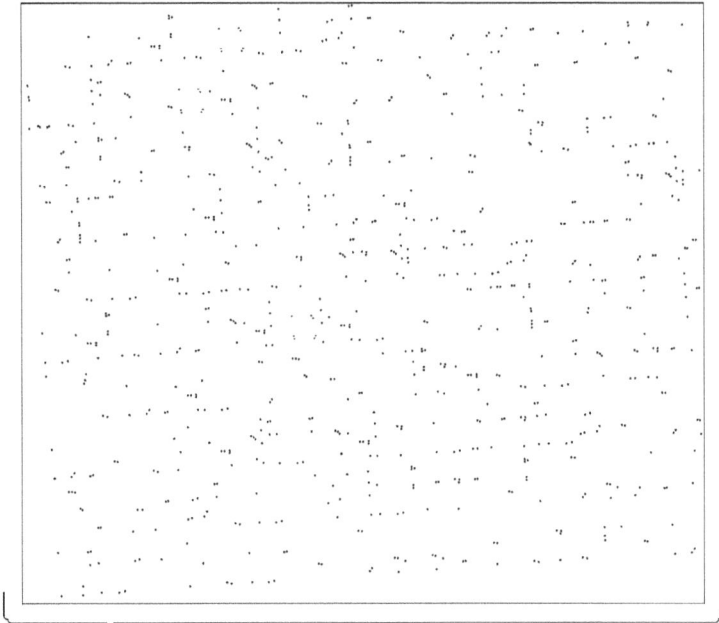

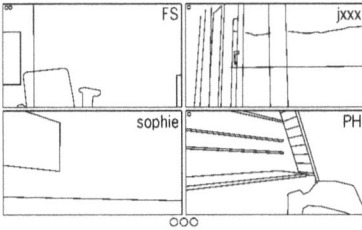

After a few weeks the window was cleaned and for a short while, only us, the round deposits, were remaining. The pression, ductus, speed, cursivity and location of where the matter was posed are dimensions that played a role in what has stayed and what went away.

We are leftovers of the drawing process. As positive letters, what else can we say or do? Lorem ipsum sam et laborrum sequaernam ut labor!

Circluding

Kym Ward feat. Possible Bodies

**This guided tour was performed on-line at *Possible Bodies Rotation
II, Imagined Mishearings* in Hangar (Barcelona, July 2017) and then
again at *Rotation III, Phenomenal 3D* in Bau (Barcelona, November
2017) with participants cutting and folding the poster reproduced on
the following pages.**[1]

Item 005: Hyperbolic Spaces

*Rolling inward enables rolling outward; the shape of life's motion
traces a hyperbolic space, swooping and fluting like the folds of a frilled
lettuce, coral reef, or a bit of crocheting.*[2]

Item 028: Circluding

*A new term, one that has been missing for a long time: "circlusion."
It denotes the antonym of penetration. It refers to the same physical
process, but from the opposite perspective. Penetration means pushing
something — a shaft or a nipple — into something else — a ring or a
tube.* Circlusion means pushing something — a ring or a tube — onto
something else — a nipple or a shaft. The ring and the tube are rendered active. *That's all there is to it.*[3]

Item 079: Gut Feminism

*The belly takes shape both from what has been ingested (from the
world), from its internal neighbors (liver, diaphragm, intestines, kidney),
and from bodily posture. This is an organ uniquely positioned, anatomically, to contain what is worldly, what is idiosyncratic, and what is visceral, and to show how such divisions are always being broken down,
remade, metabolized, circulated, intensified, and excreted. It is my concern that we have come to be astute about the body while being ignorant about anatomy and that feminism's relations to biological data
have tended to be skeptical or indifferent rather than speculative,
engaged, fascinated, surprised, enthusiastic, amused, or astonished.*[4]

Item 078: Carrier Bag Theory of Fiction

*If you haven't got something to put it in, food will escape you —
even something as uncombative and unresourceful as an oat. You put
as many as you can into your stomach while they are handy, that being
the primary container; but what about tomorrow morning when you
wake up and it's cold and raining and wouldn't it be good to have just a*

few handfuls of oats to chew on and give little Oom to make her shut up,
but how do you get more than one stomachful and one handful home?
So you get up and go to the damned soggy oat patch in the rain, and
wouldn't it be a good thing if you had something to put Baby Oo Oo in so
that you could pick the oats with both hands? A leaf a gourd a shell a
net a bag a sling a sack a bottle a pot a box a container. A holder. A
recipient.[5]

Item 80: Polyvagal Theory

The removal of threat is not the same as feeling safe.[6]

Item 81: Local Resolution

Phenomena are the ontological inseparability of agentially intra-
acting "components". That is, phenomena are ontologically primitive
relations — relations without preexisting relata. The notion of intra-
action (in contrast to the usual "interaction", which presumes the prior
existence of independent entities/relata) represents a profound concep-
tual shift. It is through specific agential intra-actions that the boundaries and properties of the "components" of phenomena become
determinate and that particular embodied concepts become meaningful. *A specific intra-action (involving a specific material configura-*
tion of the "apparatus of observation") enacts an agential cut (in con-
trast to the Cartesian cut — an inherent distinction — between subject
and object) effecting a separation between "subject" and "object". That
is, the agential cut enacts a local resolution *within the phenomenon of*
the inherent ontological indeterminacy.[7]

Notes

1. ↑ For another take on "circluding", see Possible Bodies, "Somatopologies (materials for a movie in the making)," in this book.
2. ↑ Donna J. Haraway, *Staying with the Trouble: Making Kin in the Chthulucene* (Durham: Duke University Press, 2016).
3. ↑ Bini Adamczak, "On Circlusion," *Mask Magazine*, 2016, maskmagazine.com.
4. ↑ Elizabeth A. Wilson, *Gut Feminism* (Durham: Duke University Press, 2015), 43.
5. ↑ Ursula K. Le Guin, "The Carrier Bag Theory of Fiction," in *Women of Vision: Essays by Women Writing Science Fiction*, ed. Denise Du Pont, (New York: St Martin's Press, 1988).
6. ↑ Stephen W. Porges, *The Polyvagal Theory* (New York: W. W. Norton, 2011).
7. ↑ Karen Barad, "Posthumanist Performativity: Toward an Understanding of How Matter Comes to Matter," *SIGNS* (Spring 2003): 815.

Penetration means pushing something – a shaft
or a nipple – into something else – a ring or a
tube. Circlusion means pushing something – a
ring or a tube – onto something else – a nipple
or a shaft. The ring and the tube are rendered
active. That's all there is to it.

Phenomena are the ontological
inseparability of agentially intra-acting
'components'. That is, phenomena are
ontologically primitive relations without
preexisting relata. The notion of intra-action (in
contrast to the usual 'interaction,' which presumes the
prior existence of independent entities/relata) represents
a profound conceptual shift. It is through specific agential
intra-actions that the boundaries and properties of the
'components' of phenomena become determinate and that
particular embodied concepts become meaningful. A specific intra-
action (involving a specific material configuration of the 'apparatus
of observation') enacts an agential cut (in contrast to the Cartesian
cut an inherent distinction between subject and object) effecting a
separation between 'subject' and 'object.' That is, the agential cut enacts
a local resolution within the phenomenon of the inherent ontological
indeterminacy.

Item 079: Gut Feminism

belly roils/ snaps
both from what has been
ingested (from the
world) from it's internal
neighbors' lives:
diaphragm rises/fess
kidney and liver bully
posture. This is an organ
uniquely positioned
anatomically to contain
What is worldly, what is
fabricated, and what is
needed; and to show
how such questions are
urgently being broken
metabolized, simulated,
re-routed and excreted

It is my concern that we
have come to be astute
about the body while
being ignorant about
autonomy and that
feminism's relations to
biological data have
tended to be skeptical or
ambivalent rather than
speculative, engaged,
fascinated, surprised,
astonished, amused, or
astonished

Rolling inward
enables rolling
outward; the
shape of life's
motion traces a
hyperbolic space,
swooping and
fluting like the
folds of a frilled
lettuce, coral
reef, or bit
of crocheting.

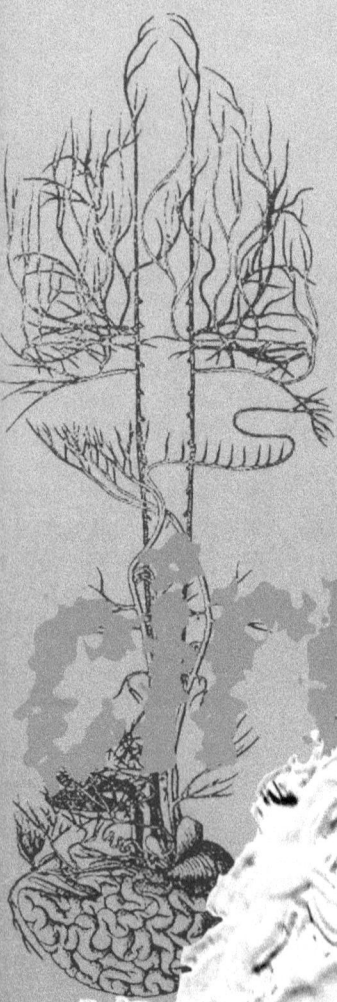

Item 078: The Carrier Bag Theory
of Fiction

If you haven't got
something to put it in,
food will escape you –
even something as
uncombative and unre-
sourceful as an oat. You
put as many as you can
into your stomach while
they are handy, that
being the primary con-
tainer; but what about
tomorrow morning
when you wake up and
it's cold and raining and
wouldn't it be good to
have just a few hand-
fuls of oats to chew on
and give little Oom to

make her shut up,
how do you get mo...
than one stomachf...
and one handful he...
So you get up and...
the damned soggy...
patch in the rain,...
wouldn't it be a go...
thing if you had so...
thing to put Baby...
Oo in so that you...
pick the oats with...
hands? A leaf a g...
a shell a net a ba...
sling a sack a bot...
put a box a conta...
A holder. A recip...

HANGAR.
ORG

Summer 2017
Copyleft: Possible Bodies / Jara Art Licence
http://possiblebodies.constantvzw.org

060: Stephen Purser: Polyglot theory
065: Karen Barad: Posthumanist Performativity,
of Japan 1923
078: Ursula K. Le Guin: The Carrier bag theory
079: Elizabeth A. Wilson: Gut Feminism 2015
nostmvgnozlm.BLOG
068: Bini Adamczak: Communism
066: Donna J. Haraway Staying with the
trouble: Making kin in the Chthulucene 2016
Cruelty – A guided tour through the mostess
bodies memory, 16-Kylin war.

MakeHuman

Jara Rocha, Femke Snelting

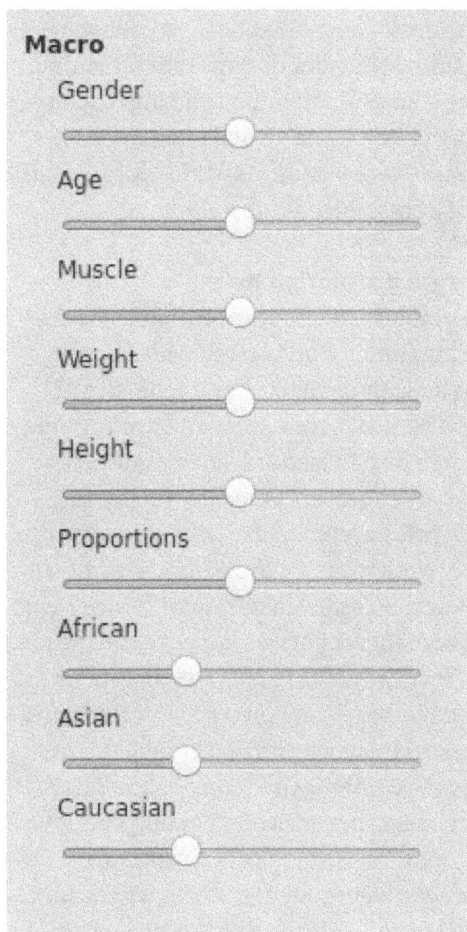

Default settings. Detail of MakeHuman's main interface (MakeHuman version 1.0.2)

MakeHuman is an Open Source software for modeling 3-dimensional humanoid characters.[1] Thinking with such a concrete software object meant to address specific entanglements of technology, representation and normativity: a potent triangle that MakeHuman sits in the middle of. But MakeHuman does not only deserve our attention due to the technological power of self-representation that it affords. As an Open Source project, it is shaped by the conditions of interrogation and transformability, guaranteed through its license. Like many other F/LOSS projects, MakeHuman is surrounded by a rich constellation of textual objects, expressed through publicly accessible source code, code-comments, bugtrackers, forums and documentation.[2] This porousness facilitated the shaping of a collective inquiry, activated through experiments, conversations and mediations.[3] In collaboration with architects, dancers, trans∗activists, design students, animators and others, we are turning MakeHuman into a thinking machine, a device to critically think along physical and virtual imaginaries. Software is culture and hence software-making is world-making. It is a means for relationalities, not a crystallized cultural end.[4]

Software: we've got a situation here

MakeHuman is "3D computer graphics middleware designed for the prototyping of photo realistic humanoids" and has gained visibility and popularity over time.[5] It is actively developed by a collective of programmers, algorithms, modelers and academics and used by amateur animators to prototype modeling, by natural history museums for creating exhibition displays, by engineers to test multi-camera systems and by game-developers for sketching bespoke characters.[6] Developers and users evidently work together to define and codify the conditions of presence for virtual bodies in MakeHuman.[7] Since each of the agents in this collective somehow operates under the Modern regime of representation, we find the software full of assumptions about the naturality of perspective-based and linear representations, the essential properties of the species and so forth. Through its curious naming the project evokes the demiurg, dreaming of "making" "humans" to resemble his own image, the deviceful naming is a reminder of how the semiotic-material secrets of life's flows are strongly linked to the ways software represents or allows so-called bodies to be represented.[8] The Modern subject, defined by the freedom to make and decide, is trained to self-construct under the narcissistic fantasy of "correct", "proper" or "accurate" representations of the self. These virtual bodies matter to us because their persistent representations cause mirror affects and effects on both sides of the screen.[9] MakeHuman is "middleware", a device in the middle: a composition machine that glues the deliriums of the "quantified self" to that of Hollywood imagery, all of it made operational through scientific anthropomorphic data and the graphic tricks of 3D-hyper-real rendering. From software development to character animation, from scientific proof to surveillance, the practices crossing through Make-Human produce images, imaginations and imaginaries that are part of a concrete and situated cultural assemblage of hetero-patriarchal positivism and humanism. Found in and fed by mainstream mediated representations, these imaginations generally align with the body stereotypes that belong to advanced capitalism and post-colonialist projections. Virtual bodies only look "normal" because they appear to fit into that complex situation.

Un-taming the whole

The signature feature of the MakeHuman interface is a set of horizon-tal sliders. For a split second, the surprising proposal to list "gender" as a continuous parameter, promises wild combinations. Could it be that MakeHuman is a place for imagining humanoids as subjects in process, as open-ended virtual figures that have not yet materialized? But the uncomfortable and yet familiar presence of physical and cul-tural properties projected to the same horizontal scale soon shatters that promise. The interface suggests that the technique of simply interpolating parameters labeled "Gender", "Age", "Muscle", "Weight", "Height", "Proportions", "Caucasian", "African" and "Asian" suffices to make any representation of the human body. The unmarked extremi-ties of the parameters are merely a way to outsource normativity to the user, who can only blindly guess the outcomes of the algorithmic calculations launched by handling the sliders. The tool invites a com-parison between "Gender" to "Weight" for example, or to slide into racial classification and "Proportions" through a similar gesture. Sub-tle and less subtle shifts in both textual and visual language hint at the trouble of maintaining the one-dimensionality of this 3D world-view: "Gender" (not "Sex") and "Weight" are labeled as singular but "Proportions" is plural; "Age" is not expressed as "Young" nor "Old", while the last slider proposes three racialized options for mixture. They appear as a matter of fact, right below other parameters, as if equal to the others, proposing racialization as a comparable and objective vector for body formation, represented as finite (and conse-quently factual) because they are named as a limited set.[10] We want to signal two things: one, that the persistent technocultural production of race is evidenced by the discretization of design elements such as the proportion of concrete bodyparts, chromatic levels of so-called skin, and racializing labels; and two, that the modeling software itself actively contributes to the maintenance of racism by reproducing representational simplifications and by performing the exclusion of diversity by means of solutionist tool operations.[11]

Further inspection reveals that even the promise of continuity and separation is based on a trick. The actual math at work reveals an extremely limited topology based on a closed system of interconnected parameters, tightening the space of these bodies through assumptions of what they are supposed to be. This risky structuration is based on reduced humanist categories of "proportionality" and "normality". Parametric design promises infinite differentiations but renders them into a mere illusion: obviously, not all physical bodies resulting from that combination would look the same, but software can make it happen. The sliders provide a machinic imagination for utilitarianised (supposedly human) compositors, conveniently covering up how they function through a mix of technical and cultural normativities. Aligning what is to be desired with the possible, they evidently mirror the binary systems of the Modern proposal for the world.[12] The point is not to "fix" these problems, quite the contrary. We experimented with replacing default values with random numbers, and other ways to intervene with the inner workings of the tool. But only when we started rewriting the interface, we could see it behave differently.[13] By renaming labels, replacing them with questions and more playful descriptions, by adding and distracting sliders, the interface became a space for narrating through the generative process of making possible bodies.

A second technique of representation at work is that of geometric modeling or polygon meshes. A mesh consolidates an always-complete collection of vertices, edges, planes and faces in order to define the topology of an individualized shape. Each face of a virtual body is a convex polygon; this is common practice in 3D computer graphics and simplifies the complexity of the calculations needed for rendering. Polygon meshes are deeply indebted to the Cartesian perspective by their need for wholeness. It results in a firm separation of first inside from outside and secondly shape or topology from surface. The particular topology of MakeHuman is informed by a rather awkward sense of chastity.[14] With all it's pride in "anatomical correctness" and high-resolution rendering, it has been decided to place genitals outside the base-body-mesh. The dis-membered body-parts are relegated to a secondary zone of the interface, together with other accessories such as hats and shoes. As a consequence, the additional add-ons so that a change in material makes them stand out, both as a potentiality for otherwise embodied otherness and as evidence of the cultural limitations to represent physical embodiment.

In MakeHuman, two different technical paradigms (parametric design and mesh-based perspective) are allied together to grow representative bodies that are renormalized within a limited and restricted field of cultivated material conditions, taming the infinite with the tricks of the "natural" and the "horizontal". It is here that we see Modern algorithms at work: sustaining the virtual by providing certain projections of the world, scaled up to the size of a powerful presence in an untouchable present. But what if the problematic understanding of these bodies being somehow human, and at the same time being made by so-called humans, is only one specific actualization emerging from an infinite array of possibilities contained in the virtual? What if we could understand the virtual as a potential generator of differentiated and differentiating possibilities? This might lead us towards mediations for many other political imaginaries.[15]

A potential for imaginations

By staging MakeHuman through a performative spectrum, the software turned into a thinking machine, confirming the latent potential of working through software objects. Sharing our lack of reverence for the overwhelming complexities of digital techniques and technologies of 3D imaging, we collectively uncovered its disclosures and played in its cracks.[16] We could see the software iterate between past and present cultural paradigms as well as between humans and non-humans. These virtual bodies co-constructed through the imagination of programmers, algorithms and animators call for otherwise embodied others that suspend the mimicking of "nature" to make room for experiences that are not directly lived, but that deeply shape life.[17]

Our persistent attention to MakeHuman being in the middle, situated in-between various digital practices of embodiment, somehow makes collaboration between perspectives possible, and pierces its own utilitarian mesh. Through strategies of "de-familiarization" the potentialities of software open up: breaking the surface is a political gesture that becomes generative, providing a topological dynamic that helps us experience the important presence of impurities in matter-culture continuums.[18] Exploring a software like MakeHuman hints at the possibility of a politics, aesthetics and ethics that is truly generative. It hints at how it is possible to provide us with endless a-Modern mestizo, an escape from representational and agential normativities, software CAN and MUST provide the material conditions for wild combinations or un-suspected renders.[19]

Notes

1. ↑ Since we wrote this text, The MakeHuman project has forked into http://makehumancommunity.org and the original website is off-line.

2. ↑ Free, Libre and Open Source Software (F/LOSS) licenses stipulate that users of the software should have the freedom to run the program for any purpose, to study how the program works, to redistribute copies and to improve the program.

3. ↑ In 2014, the association for art and media Constant organized *GenderBlending*, a worksession to look at the way 3D-imaging technologies condition social readings and imaginations of gender. The collective inquiry continued with several performative iterations and includes contributions by Rebekka Eisner, Xavier Gorgol, Martino Morandi, Phil Langley and Adva Zakai, http://genderblending.constantvzw.org.

4. ↑ http://www.makehuman.org (off-line).

5. ↑ "Makehuman is an open source 3D computer graphics software middleware designed for the prototyping of photo realistic humanoids. It is developed by a community of programmers, artists, and academics interested in 3D modeling of characters." "Makehuman," Wikipedia, accessed October 6, 2021, https://en.wikipedia.org/wiki/MakeHuman.

6. ↑ Present and past contributors to MakeHuman: http://www.makehuman.org/halloffame.php (off-line).

7. ↑ "MakeHuman," Wikipedia, accessed October 6, 2021, https://en.wikipedia.org/wiki/MakeHuman#References_and_Related_Papers

8. ↑ The Artec3 3D-scanner is sold to museums, creative labs, forensic institutions and plastic surgery clinics alike. Their collection of use-cases shows how the market of shapes circulates between bodies, cars and prosthesis. "Artec 3D scanning applications," Artec 3D, accessed October 6, 2021, http://www.artec3d.com/applications.

9. ↑ A code comment in modeling_modifiers_desc.json, a file that defines the modifications operated by the sliders, explains that "Proportions of the human features, often subjectively referred to as qualities of beauty (min is unusual, center position is average and max is idealistic proportions)." https://bitbucket.org/MakeHuman/makehuman (version 1.0.2).

10. ↑ humanmodifierclass.py, a file that holds the various software-classes to define body shapes, limits the "EthnicModifier(MacroModifier) class" to three racial parameters, together always making up a complete set: # We assume there to be only 3 ethnic modifiers. self._defaultValue = 1.0/3" https://bitbucket.org/MakeHuman/makehuman (version 1.0.2).

11. ↑ humanmodifierclass.py, a file that holds the various software-classes to define body shapes, limits the "EthnicModifier(MacroModifier) class" to three racial parameters, together always making up a complete set: # We assume there to be only 3 ethnic modifiers. self._defaultValue = 1.0/3" https://bitbucket.org/MakeHuman/makehuman (version 1.0.2).

12. ↑ In response to a user suggesting to make the sliders more explicit ("It really does not really make any sense for a character to be anything other then 100% male or female, but than again its more appearance based than actual sex."), developer Manuel Bastioni responds that it is "not easy: For example, weight = 0.5 is not a fixed value. It depends by the age, the gender, the percentage of muscle and fat, and the height. If you are making an adult giant, 8 ft, fully muscular, your 0.5 weight is X. [...] In other words, it's not linear." Makehuman, http://bugtracker.makehumancommunity.org/issues/489.

13. ↑ MakeHuman is developed in Python, a programming language that is relatively accessible for non-technical users and does not require compilation after changes to the program are made.

14. ↑ When the program starts up, a warning message is displayed that "MakeHuman is a character creation suite. It is designed for making anatomically correct humans. Parts of this program may contain nudity. Do you want to proceed?"

15. ↑ The trans∗-working field of all mediations is a profanation of sacred and natural bodies (of virtuality and of flesh). It evidences the fact of them being technological constructions.

16. ↑ Here we refer to Agamben's proposal for "profanation": "To profane means to open the possibility of a special form of negligence, which ignores separation or, rather, puts it to a particular use." Giorgio Agamben, *Profanations* (New York: Zone Books, 2007), 73.

17. ↑ "The ergonomic design of interactive media has left behind the algorithmic 'stuff' of computation by burying information processing in the background of perception and embedding it deep within objects." Luciana Parisi, *Contagious Architecture: Computation, Aesthetics, and Space* (Cambridge MA: MIT Press, 2013).

18. ↑ Breaking and piercing the mesh are gestures that in "This topological dynamic reverberates with QFT processes [...] in a process of intra-active becoming, of reconfiguring and trans-forming oneself in the self's multiple and dispersive sense of it-self where the self is intrinsically a nonself." Karen Barad, "TransMaterialities: trans∗/Matter/Realities and Queer Political Imaginings," *GLQ* 21, nos. 2-3 (June 2015): 387-422.

19. ↑ "xperiments in virtuality - explorations of possible trans∗formations- are integral to each and every (ongoing) be(coming)." Barad, *TransMaterialities*.

Information for Users

Jara Rocha, Femke Snelting

This information leaflet accompanied a banner for a course, *Somatic Design*, which depicted five 3D-generated humanoid representations. The pamphlet circulated in the hallways of an art school in Barcelona, May 2015.

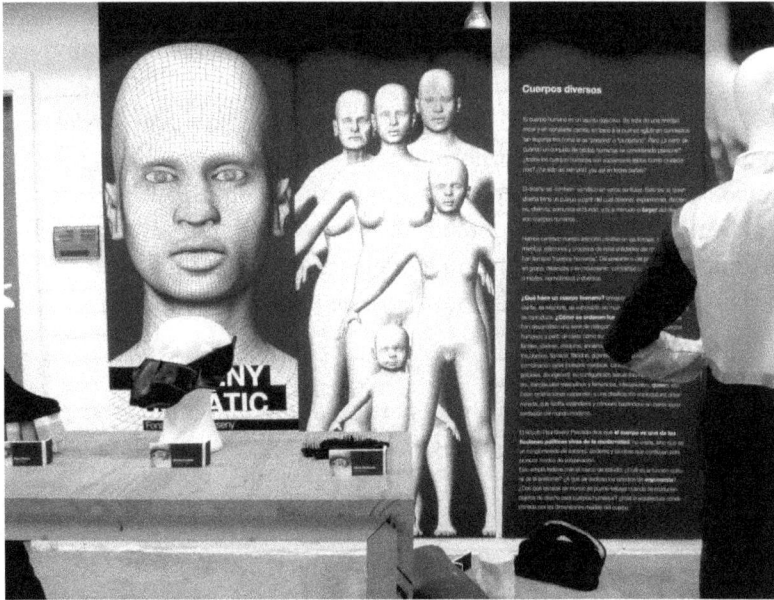

Please read carefully. This leaflet contains important information:

- Save this leaflet, it might be useful in other circumstances.
- If you have additional questions, discuss with your colleagues.
- If you experience a worsening of your condition, document and publish.
- If you experience any of the side-effects described in this leaflet or you experience additional side-effects not described in this leaflet, report a bug.
- See under 5. to find out if you are specifically at risk.

1. What is this image?

The image is circa 80 cm wide and 250 cm high, printed on a high resolution inkjet printer.It accompanies a display of results from the course Somatic Design — Fonaments del Disseny I (2014-2015) — and can be found in the hallway of Bau, Design College of Barcelona, May 12-18, 2015. The image consists of five 3D-generated humanoid representations, depicted as wireframe textures on a white contour, placed on a blood-red background. The 3D-generated humanoid representations are depicted at nearly life-size, without clothes and holding the same body posture.

The software used to generate this image is MakeHuman, an "open source tool for making 3D-characters".

The perspective used is orthogonal, the figures appear stacked upon each other. Height is normalized: the figure representing a grown-up male is larger than the female, the older female figure is smaller than the younger female.The genitals of the largest male figure and elder female figure are hidden; the genitals of the adult female figure are only half-shown; the genitals of the children are shown frontally.

2. Important information about 3D-generated humanoid representations:

- There is an illusionary trick at work related to the resolution of the image. 3D-generated images might appear hyper-real, but are generated from a crude underlying structure.
- 3D-generated imagery has a particular way of dealing with inside and outside. The "mesh" that is depicted here as a wireframe, necessitates a binary division between inside and outside, between flesh and skin.
- Software for generating 3D humanoid representations is parametric. This means that its space of possibilities is predefined.
- The nature of the algorithms used for generating these representations, has an effect on the nature of the representation itself.
- 3D-generated humanoid representations often depart from a fundamentally narcissistic structure.

- These 3D-generated images are aligned with a humanist cultural paradigm, otherwise known as The Modern Project. They are not isolated from this paradigm, but are evidence of an epidemic.
- The Modern Project produces a desire for an ecstasy of the real.

3. Before engaging with humanoid representations:

- Remember that viewing images has always an effect. In this leaflet engagement is used rather than seeing or looking. It is not possible to view without being transformed.
- Representations are made by a collective of humans and non-humans. Here, algorithms and tools are co-designing.
- Scientific data suggests perfection through averaging. An average is the result of a mathematical calculation and results in hypochondria.

4. In case humanoid representations are grouped:

- What is placed in the foreground and what is placed in the background matters. If bodies are ordered by size and age (for example smaller and younger in the foreground, larger and older in background), a hierarchy is suggested that might not be there.
- Size matters. The correlation between age, gender and size is usually not corresponding to the average.
- Nuclear families are not the norm. The representation of gender and age, as well as the number of bodies depicted, is always a decision and never an accident.
- The depiction of figures with a variety of racial physiological features matters. Even if this group is not all Caucasian, There is no mestizo in the image. The reality of hybridization is more complex.
- The lack of resemblance to how people physically relate in daily life, matters. Bodies are not usually stacked that closely, nor positioned behind each other frontally, neither holding all the exact same body posture.

- The represented space for relational possibilities can be unnecessarily limited. For example: if in a group only one male is depicted, it is assumed that this body will relate to the others in a hetero-patriarchal manner.

5. Counter-indications:

Take care if you are concerned by the over-representation of nuclear families. Be especially careful with this type of image if:

- You have (or belong to) a family.
- You are pregnant or lactating.
- You feel traumatized by hetero-patriarchal, capitalist or religious institutions.
- Your body type does not fit.
- You think another world is possible.
- Your unconscious shines.

6. How should I engage with this image?

Approach these images with care, especially when you are alone. It is useful to discuss your impressions and intuitions with colleagues.* Try out various ways of critically engaging with the representation.

- Measure yourself and your colleagues against this representation.
- Try decolonial perspectives.
- Ask questions about the ordering of figures, what is made visible and what is left out.
- Ask why these humanoids do not have any (pubic) hair.
- Problematize the parametric nature of these images: What is their space of possibilities?

Be aware of your desire apparatus.

7. Interactions with other images:

These images are part of an ecosystem: they generally align with gen-der-stereo-types and neoliberal post-colonialist imagery, found in mainstream media. They might look "normal" just because they seem to fit this paradigm.

Pay attention to the hallucinatory effect of repetition.

8. What to avoid while engaging with this image:

Avoid trusting this image as a representation of your species. The pseudo-scientific atmosphere it creates is an illusion, and constructed for a reason. Do not compare yourself with these representations.

9. What are the most common side effects of engaging with humanoid representations:

- Vertigo and dis-orientation
- A general feeling of not belonging
- Anger, frustration
- Insomnia, confusion
- Nausea
- Speechlessness
- An agitation of life conditions
- It may increase thinking or extreme questioning

10. In case of overdose:

In case of overdose, a false sense of inclusion might be experienced. Apply at least three of the methods described under 5. Repeat if necessary until the condition ameliorates.

INFORMATION FOR USERS

INFORMATION LEAFLET TO ACCOMPANY A BANNER FOR *SOMATIC DESIGN*, DEPICTING FIVE 3D-GENERATED HUMANOID REPRESENTATIONS

Please read carefully. This leaflet contains important information:

- Save this leaflet, it might be useful in other circumstances.
- If you have additional questions, discuss with your colleagues.
- If you experience a worsening of your condition, document and publish.
- If you experience any of the side-effects described in this leaflet or you experience additional side-effects not described in this leaflet, report a bug.
- See under 4 to find out if you are specifically at risk.

1. What is this image?

The image is circa 80 cm wide and 250 cm high, printed on a high resolution inkjet printer.
It accompanies a display of results from the course *Somatic Design – Fonaments del Disseny I* (2014-2015) and can be found in the hallway of Bau, Design College of Barcelona, May 12–18, 2015. The image consists of five 3D-generated humanoid representations, depicted as wireframe textures on a white contour, placed on a blood-red background. The 3D-generated humanoid representations are depicted on nearly life-size, without clothes and holding the same body posture.

The software used to generate this image is MakeHuman, an 'open source tool for making 3D-characters'.

The perspective used is orthogonal, the figures appear stacked upon each other. Height is normalized: the figure representing a grown-up male is larger than the female, the older female figure is smaller than the younger female. The genitals of the largest male figure and elder female figure are hidden; the genitals of the adult female figure are only half-shown; the genitals of the childeren are shown frontally.

2. Important information about 3D-generated humanoid representations:

- There is an illusionary trick at work related to the resolution of the image. 3D-generated images might appear hyper-real, but are generated from a crude underlying structure.
- 3D-generated imagery has a particular way of dealing with inside and outside. The 'mesh' that is depicted here as a wireframe, necessitates a binary division between inside and outside, between flesh and skin.
- Software for generating 3D humanoid representations is parametric. This means that its space of possibilities is pre-defined.
- The nature of the algorithms used for generating these representations, has an effect on the nature of the representa-tion itself.
- 3D-generated humanoid representa-tions often depart from a fundamentally narcissistic structure.
- These 3D-generated images are aligned with a humanist cultural paradigm, otherwise known as The Modern Project. They are not isolated from this paradigm, but are evidence of an epidemy.
- The Modern Project produces a desire for an ecstasy of the real.

3.1 Before engaging with humanoid representations:

- Remember that viewing images has always an effect. In this leaflet engagement is used rather than *seeing* or *looking*. It is not possible to view without being transformed.
- Representations are made by a collec-tive of humans and non-humans. Here, algorithms and tools are co-designing.
- Scientific data suggests perfection through averaging. An average is the result of a mathematical calculation and results in hypochondria.

3.2 In case humanoid representations are grouped:

- What is placed in the foreground and what is placed in the background matters. If bodies are ordered by size and age (for example smaller and younger in the foreground, larger and older in background), a hierarchy is suggested that might not be there.
- Size matters. The correlation between age, gender and size is usually not corresponding to the average.
- Nuclear families are not the norm. The representation of gender and age, as well as the number of bodies depicted, is always a decision and never an accident.
- The depiction of figures with a variety of racial physiological features matters. Even if this group is not all caucasian, There is no mestiza in the image. The reality of hybridisation is more complex.
- The lack of resemblance to how people physically relate in daily life, matters. Bodies are not usually stacked that closely, nor positioned behind each other frontally, neither holding all the exact same body posture.
- The represented space for relational possibilities can be unnecessarily limited. For example: If is a group only one male is depicted, It is assumed that this body will relate to the others in a hetero-patriarchal manner.

4. Counter-Indications:

Be especially careful with this type of image if:

- You have (or belong to) a family.
- You are pregnant or lactating.
- You feel traumatized by hetero-patriarchal, capitalist or religious institutions.
- Your body type does not fit.
- You think another world is possible.
- Your unconscious shines.

Take care if you are concerned by the over-representation of nuclear families.

5. How should I engage with this image?

Approach these images with care, especially when you are alone. It is useful to discuss your impressions and intuitions with colleagues.

- Try out various ways of critically engaging with the representation.
- Measure yourself and your colleagues against this representation.
- Try decolonial perspectives.
- Ask questions about the ordering of figures, what is made visible and what is left out.
- Ask why these humanoids do not have any (pubic) hair.
- Problematize the parametric nature of these images: What is their space of possibilities?

Be aware of your desire apparatus.

6. Interactions with other images:

These images are part of an ecosystem: they generally align with gender-stereo-types and neoliberal post-colonialist imagery, found in mainstream media. They might look 'normal' just because they seem to fit.

Pay attention to the hallucinatory effect of repetition.

7. What to avoid while engaging with this image:

Avoid trusting this image as a representation of your species. The pseudo-scientific atmosphere it creates is an illusion, and constructed for a reason. Do not compare yourself with these representations.

8. What are the most common side effects of engaging with humanoid representations:

- Vertigo and dis-orientation
- A general feeling of not belonging
- Anger, frustration
- Insomnia, confusion
- Nausea
- Speechlessness
- An agitation of life conditions
- It may increase thinking or extreme questioning

9. In case of overdose:

In case of overdose, a false sense of inclusion might be experienced. Apply at least three of the methods described under 5. Repeat if necessary until the condition ameliorates.

This leaflet provides basic information about engaging with 3D-generated humanoid representations.

After reading, it is not uncommon that an initially passive reception transforms into another level of engagement. You might experience a desire to generate action and political work vis-a-vis humanoid and other representations. In this case, it is advisable that you approach your condition collectively.

Signs of Clandestine Disorder:
The continuous aftermath of 3D-computationalism

Endured Instances of Relation

Romi Ron Morrison in conversation with Jara Rocha and Femke Snelting

After listening to your talk *The forgotten past of black computational thought*,[1] we would like to ask you about your specific understanding of what "difference without separation" could mean. We are trying to think about separation and difference specifically in relation to volumetric computational processes that de-flatten or re-flatten, model, capture, track and so forth.

I think entanglement is the word.

For me, your question seems to recursively return to this. Entanglement implies a relation. Perhaps one that evades or overdetermines what cannot presently be grasped but nonetheless, a relation. Entanglement is helpful for me to think through because it doesn't resolve into an easy self contained knowability, but it also doesn't mask itself within the complete opacity of being unknowable to the extent of any totality. Rather, entanglement moves towards a question of "how" and "what if". It refuses the punctuation of a period to give space for what follows. It is something we must work with outside of pursuits of resolution, and each attempt is one that strives for a better understanding of the richness of the relation. To engage entanglement in this way is a practice of endurance.

Thinking about the questions that you have asked to start this conversation, difference without separability is invested in these spaces of entanglement, or perhaps what Glissant would call a poetics of duration, of relation. This phrase "difference without separability" comes from Denise Ferreira da Silva's work. In her article, "On Difference Without Separability", da Silva gives a brief history of modern thought through Descartes, Newton, Kant, Cuvier, Boas and Foucault. She traces the ways that these "modern texts" scientifically image The World as an "ordered whole composed of separate parts relating through the mediation of constant units of measurement and/or a limiting violent force".[2] This separability is a constitutive component for ushering in modernity by which difference is rendered as fixed and irreconcilable. This negation built upon the overrepresentation of the human as Man, is what upholds the human (body as sovereign property) as a moral figure that necessitates the edgeless violence of enslavement and genocide on those deemed nonhuman or partially human (body as flesh). This separability is a crucial modern text that fixes the present world in a scene of constant reenactment of these violences though the name of the violence has shifted and is proclaimed as national security, sovereignty, austerity, structural adjustment, sanctity of the family, or freedom. In *The forgotten past of black computational thought*, I speak of an operating system overdetermined by anti-black violence regardless of who the programmer is, I am speaking to the repetition of this logic of separability that is constituted through a justification of violence.

Separability is built upon a kind of racial technoscience. It severs the possibility of relation and masks entanglement in pursuit of the pure. There are only rounded decimals here, they always terminate. Thinking about your interest in "bodies" and the ways that they are rendered and constituted through volumetric digital technologies, this emphasis on separability is germane, as possible bodies become captured into standard fixed units of difference.

In hegemonic applications of computation, we see that separation is supposed to function as a neutral, necessary, efficient gesture. Do you think this is how anti-blackness ends up in the bowels of computation? Is it already prefigured in the binary "nature" of computing, not just as a technical basis, but also as an ethics a politics and material culture? Is separation where the coerciveness of computation stems from? And if computation is inherently anti-black, does it make sense to ask it to engage with other lives and relationalities, such as fair algorithms, data justice and infrastructures of care?

I return to this separability because it seems so central for understanding and rethinking both the violences and possibilities for computation. In my prior talk that you referenced, I am trying to make a connection between separability in the da Silvian sense and what David Golumbia calls "computationalism". Golumbia makes a distinction between computers and computationalism. For him computationalism "is the view that not just human minds are computers but that mind itself must be a computer—that our notion of intellect is, at bottom, identical with abstract computation".[3] Computationalism understands cognition itself as inherently a computing process, and by extension, all matters of phenomena in the world can be understood as a function of computation. Thinking about computationalism rather than computing or computation potentially frees the latter from the violences of the former and opens some space for experimentation and reimagining. Computationalism inherits the violences of the modern text that da Silva details. Its central episteme upheld by irreconcilably fixed difference, universal measurements, and separation continues largely undisturbed.

How to think about messiness in relation to possible forms of computation? Flesh, complexity and mess are also already-with computation, not before or after data, but somehow simultaneous and constituent of computation and constituent of mess in reciprocity. How could computation and flesh together constitute more livable messes, if at all?

Sketching the shared contours between modernity, and its dependence on black and native violence, and to call it "computationalism" perhaps allows for computing to return to a much more expansive capacity that doesn't always require such violence. This is where I'm interested in speculation and in particular speculative histories, presents, and futures of computation that come out of the political, poetic, and erotic practices of blackness and fugitive fungibility. This thinking thrives in relationship to the work of black queer, trans, feminist scholars and artists such as Hortense Spillers, Sylvia Wynter, C. Riley Snorton, Tiffany Lethabo King, Tina Campt, Saidiya Hartman, Katherine McKittrick, and Marquis Bey. Rather than taking up the body as a site of the liberal human subject imbued with agency, ownership, and stability, these scholars theorize through the flesh and fungibility of blackness. Flesh is distinguished from the body as a result of the unimaginable violence wrought on black people in making them property, unfree laborers, and fungible sites of death, expansion, desire, sensuousness, and commodity. Spillers and King in particular write about the ways in which Black people under capture, conquest, and enslavement were made fungible. They were made into constantly exchangeable resources able to malleably stand in for any needs white colonizers could imagine. While fungibility is born from and determined by continuous violence, Snorton also notices the simultaneous life and possibility even in the shadow of such death. For Snorton fugitive fungibility marks a space of indeterminacy and possibility, which might open other ways of being outside the trappings of the human. This fleshy fungibility is a porous space to inhabit that exists in shared relations to land and other nonhuman and extrahuman others. It is a relation of entanglement. From this place I hope to speculate on different forms of computing that thrive in indeterminacy and work from an ethical relationship of entanglement.

Thinking computation from this place works from the assumptions that computation cannot be done away with as a means of addressing violence. It understands that computation is a method, practice, ideology, and episteme. And in its most hegemonic understanding is a very limited form of discourse. As many of the theorists above hold no romances about the extent and saturation of anti-black violence in the modern world, they also tend to the possibilities of life and living that extend beyond that violence. While violence cannot be ignored, it also doesn't overdetermine life to the extent of rendering it abject and wholly without. I believe it is possible to contend with the violences of computation while simultaneously lingering in the vitality of the flesh. To think and practice computing otherwise as technologies of the flesh that thrive within indeterminacy and interdependency. This is what informs where I think we might look to recover some of these forms. Within my work I look at practices of computation that live in the poetics, politics, erotics, and movements of blackness.

Through your studies of the legacies of code, you ask: What if computation engaged with indexing different zones of life, facilitated relationalities other than those of capitalist anti-blackness? Could you say more about the kind of computation this would generate, because you seem to call into question most of all that which is indexed and who is indexing, rather than indexing as a problem in and of itself? The question could also be formulated like this: is there space for attending to volumes technically in their singularity, while not reproducing the exclusions that the very techniques of measuring carry? Or, are there other uses of volumetric techniques that apply separation and indexing, while disassembling those practices from the episteme of exclusion?

As you referenced earlier, my interests in fugitive fungibility informs how I have been thinking about indexing and the database as a potential space to make connections and practice a kind of endured proximity by which we are in relation to that which we index. That we can be in a fungible relationship through porosity. That entanglement is allowed to exist and can be seen as a source for ethical encounter. I suppose this would drastically change how we consider indexing and what we consider indexing to be. Within current hegemonic practices of data capture and indexing the world through measuring, there are certain paradigms that need to be challenged. For me these primarily stem from separability by which measurement simultaneously fixes difference as stable and as irreconcilable. Rather, I believe indexing can hold a different potential when deracinated from this episteme of separability. Instead I think of indexing as a way of accounting for an instance of something. And that because of its shared relations it evades static standardization and is instead in flux and changing. I suppose this gives more texture to the ways that I think about entanglement. Or to be more direct, I believe the benefits of indexing are temporally bounded. They are not absolute nor axiomatic. But I believe indexing can also serve to better emphasize the multiple relations between things in a much more robust way than simply the observable measured differences that scientific rationality often privileges. This form of indexing is malleable and contextual, it depends on the one indexing, the method, and on that which is indexed. Its endured proximity doesn't seek to remove complications through the rhetoric of universality or transparency, but is invested in the particular and chronic.

Computation and life ("bodies", spaces, relationalities) are already entangled in so many ways; they are mutually constituent, for example the category of life wouldn't exist without a whole apparatus of segmentation producing it as different from the non-living. To us it feels urgent to think with and towards computing-otherwise rather than to side with the uncomputable or to count on that which escapes calculation. What would it mean to critique math and quantification in their Modern shape, by calling for other logics instead?

In earlier writing, I have returned to theorist and filmmaker Trinh T. Minh-ha's practice of *speaking nearby* to illustrate this relationship.[4] In an interview with Nancy N. Chen for the *Visual Anthropology Review*, Minh-ha elaborates further: "In other words, a speaking that does not objectify, does not point to an object as if it is distant from the speaking subject or absent from the speaking place. A speaking that reflects on itself and can come very close to a subject without, however, seizing or claiming it. A speaking in brief, whose closures are only moments of transition opening up to other possible moments of transition".[5] I believe this could be an opening potential for indexing and the database, as a temporal marker of an instance of something in relation. What it tells us is not data about the essence of a fixed object, but of something caught in flux that we are in relation to.

I also think this is a place where different practices of computation can be speculated on. To be able engage this type of indexed entanglement, it opens questions of method or protocol. It requires practice. More and more, I stick with computation to describe some of this complexity for a few reasons. The first is in refusing to relinquish computation as an already closed system that no longer requires definition. The second is in acknowledging the economic, cultural, imaginative, and disciplinary power that computation presently holds. And lastly, to speculate on the unique capacity of computation to contend with complex variables and their relationship to flux and modulation.

Speaking on this capacity, Édouard Glissant writes about the trappings and potential that the computer holds towards poetics. In his text, *Poetics of Relation*, Glissant briefly discusses computation and how it differs from poetry. On this he writes, "Accident that is not the result of chance is natural to poems, whereas it is the consummate vice (the "virus") of any self-enclosed system, such as the computer. The poet's truth is also the desired truth of the other, whereas, precisely, the truth of a computer system is closed back upon its own sufficient logic. Moreover, every conclusion reached by such a system has been inscribed in the original data, whereas poetics open onto unpredictable and unheard of things."[6] Glissant contrasts computation and poetry focusing on the closed, controlled, and binary character of computationalism. He understands it as a mechanism of separability. However, the potential for the computer when working outside of computationalism is not foreclosed. Just a few pages later he writes, "The computer, on the other hand, seems to be the privileged instrument of someone wanting to "follow" any Whole whose variants multiply vetiginously. It is useful for suggesting what is stable within the unstable. Therefore, though it does not create poetry, it can 'show the way' to a poetics."[7]

Because computation is able to contend with complex multiplicity Glissant leaves it open as a wayfinder towards a poetics. He makes a slight but crucial distinction that computation is useful for suggesting what is stable within the unstable. He doesn't state that computing itself creates stability or static fixed variables, but instead is able to suggest stability as an open and incomplete instance within a field of instability. While his first quote indexes some of the trappings of computation as a closed logic, he follows it by hinting at the possibility for computation to move through the complexities of entanglement. Perhaps at best, computation in this sense can hold the tension of indeterminacy without either becoming paralyzed or reducing the complexity of the Whole into predictable calculable units. Within this slight shift in language, computation is nudged open. It is made porous again and moves towards the direction of a poetics. Perhaps then this porousness can allow for finding a poetics of space within volumetric capture, by underlining the stable and unstable within computation, and resituated computation as a manner and mode of engaging the entanglement between those two poles. It is a practice of "showing the way" to a relation. Both bodies and space in this mode of computation hold a certain openness. They cannot completely be foreclosed as inherently separable parts.

We wondered about the voluminosity of "bodies" but also of entanglement, and how to pay attention to it. Reading Denise Fereirra da Silva's email conversation with Arjuna Neuman about her use of "Deep Implicancy" rather than "entanglement", we were struck by the relation between spatiality and separation she brings up: "Deep Implicancy is an attempt to move away from how separation informs the notion of entanglement. Quantum physicists have chosen the term entanglement precisely because their starting point is particles (that is, bodies), which are by definition separate in space."[8]

So what if the spaces of entanglement provide a semiotic-material arena for cohabiting with and practicing 3D computation-otherwise? Could "Deep Implicancy" be where computing otherwise already happens, by means of speculation, indeterminacy and possibility located beyond, or below perhaps, normed actions like capturing, modeling or tracking that are all so complicit with the making of fungibility?

So this question of Deep Implicancy is interesting. I think in reading through da Silva and Neumann's email exchanges, I have a sense of the difference that she is trying to draw between entanglement and its inherent dependence on a kind of separability, because of its embedded focus on particles inherited from physics. Even things such as quantum entanglement or nonlocality, are still built from some kind of separability. I think that is an important distinction and contribution which breaks open some of my earlier thoughts on entanglement. That being said, I'm not sure I understand Deep Implicancy beyond the ways that it complicates the inherent separability within entanglement. It makes me want to ask, how does Deep Implicancy account for or contend with difference? It seems that there would still need to be room for variation or modulation. Perhaps even modulation and distance can become the language through which to speak to fluctuations, changes, variations, and instances within a

dynamic implicancy. Because then we are able to account for difference without flattening it to an equivalence or commensurability.
This thinking on modulation and difference is very much informed by
Kara Keeling's work in *Queer Times Black Futures*,[9] and Abdoumaliq
Simone's work in *Improvised Lives: Rhythms of Endurance in an Urban
South*.[10] In her discussion of James A. Snead's work on Black culture
and repetition,[11] Keeling makes connections to the computational
practice of modulation and incommensurability. Evoking Snead, she
states, "repetition means that the thing circulates (exactly in the manner of any flow, including capital flows) there in an equilibrium". The
"thing (the ritual, the dance, the beat) is there for you to pick up when
you come back to get it". She argues that this repetition and the ability
to return rather than progress allows for a kind of cultural coverage
that builds spaces for the unpredictable, errant, and accidental to
happen. Keeling sees this practice as a mechanism of modulation, a
mode of social and cultural continuity, which does not rely upon com-
mensuration. Instead, it makes "incommensurability" into a relation.
Perhaps this incommensurability, the impossibility of neat resolve
can provide a helpful language to engage Deep Implicancy and its
relationship to difference.

**The episteme of Modern technosciences classifies "bodies" as
entities that occupy the dimensions of space and time at a certain
scale, with a certain density, at a certain speed, etc. It is complicit
with productivist, segregating, extractivist and deadly aims when
calculating volumes of so-called bodies and their surroundings. But
maybe such displacements, dimensional and material conditions,
could also be of use for a disobedient rearranging of so-called bodies? How to think with possible forms of computation that do not
leave its oppressions in place?**

Simone picks up this relation of incommensurability and
stretches it to describe the movements, motions, calculations, and
alterations of bodies as they converge and depart in space. Simone
describes these bodies as "technical forces" that "speak, spit, stomp,
fuck, gesture, lunge, or hover". His understanding of space is constructed through these rhythms of endurance that bodies undertake
in a constant renegotiation towards "a liveliness of things in general".
For Simone, "endurance also entails the actions of bodies indifferent
to their own coherence, where bodies proliferate a churning that
staves off death in their extension toward a liveliness of things in general, and where bodies become a transversal technology, as gesture,

sex, gathering, and circulation operate as techniques of prolonging".[12] His writings on bodies as transversal technologies is really intriguing, in that they are always intersecting, crossing, and circulating. In doing so, it creates the spaces that they momentarily inhabit. The space does not precede the bodies. It is not a container in this analysis but is constructed through the circuitous gestures, gatherings, and sex of bodies churning together in incommensurability. Similarly, to Keeling's focus on repetition Simone offers us a musical lexicon of rhythm, refrain and pulse to find stabilizing moments that thrive in response to risk and incalculability. For Simone the refrain works as this stabilizing repetition that creates "contexts of operation that cannot be stabilized". Again, space for Simone is dependent and created through these undulating intersections of bodies that enact open modulating refrains. This works against easy practices of tracking or capturing, that volumetrically rendered spaces require, as it exceeds any preemptive containment. Space for Simone is not predetermined but is interdependent. More importantly, it is interdependent on the relations of bodies that evade stable categorization or coherence. Instead these relations are constantly modulating and shifting. Perhaps most beautifully, Simone articulates these intersecting modulations as care. On this he writes:

> For the intersections among spiraling trajectories are a matter of care[13], inexplicable care, rogue care, care on the run, a tending not to people or by people, but a care that precedes them. It is a care that makes it possible for residents to navigate the need to submit and exceed, submerge themselves into a darkness in which they are submerged but to read its textures, its tissues, to see something that cannot be seen. It enables them to experience the operations of a sociality besides, right next to the glaring strictures of their obligations, expulsions, and exploitation, something that enables endurance, not necessarily their own endurance as human subjects, but the endurance of care indifferent to whatever or whoever it embraces. This is a process that entails both composition and refusal.[14]

Care here seems to emerge as an ethic void of preconditions. It simply is because it must be. It is a practice of endurance outright. One that enables fugitive flights, the promise of continued evasion, and a relation beyond commensurable equivalences. Perhaps this gives us more texture for what a Deep Implicancy can offer, no longer entangled, but stomping, speaking, and spitting in a space made through care without preconditions, indifferent to quantification.

Notes

1. ↑ Romi Ron Morrison, "Speaking Nearby: The forgotten past of computational thought", paper presented at EASSST/4S *Conference: Crafting Critical Methodologies in Computing: Theories, Practices and Future Directions,* Prague, August 18-21, 2020.

2. ↑ Denise Ferreira da Silva, "On Difference Without Separability," in *Incerteza Viva: 32nd Bienal de São Paulo,* ed. Jochen Volz and Júlia Rebouças (Sao Paulo, Ministry of Culture, Bienal and Itaú, 2016), 57-58.

3. ↑ David Golumbia, *The Cultural Logic of Computation* (Cambridge MA: Harvard University Press, 2009), 7.

4. ↑ Nancy N. Chen, "Speaking Nearby: A Conversation with Trinh T. Minnh-ha," *Visual Anthropology Review* 8, no. 1 (Sping 1992): 87.

5. ↑ Chen, "Speaking Nearby," 87.

6. ↑ Édouard Glissant, *Poetics of Relation* (University of Michigan Press, 1997), 82.

7. ↑ Édouard, *Poetics of Relation,* 84.

8. ↑ Email correspondence between Arjuna Neuman and Denise Ferreira da Silva, 2017-2018 https://www.thes howroom.org/system/files/062020/5 ef3716252712a038b005fbc/original/e mail_correspondence_AN_DFDS.pd f?1605089604.

9. ↑ Kara Keeling, *Queer Times Black Futures* (New York: New York University Press, 2019).

10. ↑ AbdouMaliq Simone, *Improvised Lives: Rhythms of Endurance in an Urban South* (Cambridge: Polity Press, 2019).

11. ↑ James A. Snead, "On Repetition in Black Culture," *African American Review* 50, no. 4 (Winter 2017).

12. ↑ Simone, *Improvised Lives.*

13. ↑ María Puig de la Bellacasa, *Matters of Care: Speculative Ethics in More than Human Worlds* (University of Minnesota Press, 2017).

14. ↑ AbdouMaliq Simone, *Improvised Lives.*

The Industrial Continuum of 3D

Jara Rocha, Femke Snelting

The Invention of the Continuum

Whether it is cultural heritage, archaeological sites or the natural world, his personal mission is to build technologies that help explore the world and the disappearing things around us. The engineer and entrepreneur aims an arsenal of synchronized cameras at a caged rhinoceros, and explains: "In the end, you will be able to stand next to the rhino, look into the animal's eye and this creates an emotional connection that is beyond what you can get from a flat video or photograph. The ultimate application will be, to bring the rhino to everyone."[1]

3D scanning a specimen of the near-extinct Sumatran rhinoceros as an act of conservation turns the 6th extinction into a spectacle. As a last-minute techno-fix, it renders "the ultimate application" that is available for everyone at home, while the chain of operations it participates in technically contributes to extinction itself. Capturing the rhinoceros depends on mineral extraction and the consumption of turbo-computing, and also continues to trust in the control over time via techno-solutionist means such as volumetric capture and the wicked dream of re-animation cloaked as digital preservation.

The industrial continuum of 3D is a sociotechnical phenomenon that can be observed when volumetric techniques and technologies flow between industries such as biomedical imaging, wild life conservation, border patrolling and Hollywood computer graphics. Its fluency is based on an intricate paradox: the continuum moves smoothly between distinct, different or even mutually exclusive fields of application, but leaves very little space for radical experiments and surprise combinations. This text is an attempt to show how the consistent contradiction is established, to see the way power gathers around it, to get closer to what drives the circulation of industrial 3D and to describe what is settled as a result. We end with a list of possible techniques, paradigms and procedures for "computing otherwise", wondering what other worldings could be imagined.[2]

We have named this continuum *industrial* because its flows are driven by the rolling wheels of extractive *patriarchocolonial* capital. Think of the convenient merging of calculations for building and for logistics in 3D model-based architectural processes such as Building Information Modeling (BIM).[3] Or think of the efficacy of scanning the underground for extractable resources with the help of technologies first developed for brain surgery. Legitimated areas of research spill into management zones with oppressing practices, and in the entrepreneurial eyes of old Modern scientists, the research glitters with startup hunger, impatient to serve the cloudy kingdom of GAFAM.[4] The continuum continuously expands, scales up and down, connecting developed arenas with others to be explored and extracted. Volumetric scanning, tracking and modeling obviously share some of the underlying principles with neighboring hyper-computational environments, such as machine learning or computer vision,[5] but in three-dimensional operations, the industrial continuum intensifies due to their supercharged relationship to space and time.[6]

By referring to this phenomenon as a "continuum", we want to foreground how rather than prioritizing specificity, it thrives on *fabricating similarities* between situations. Its agility convokes a type of space-time that is both fast and ubiquitous, while relegating the implications of its industrial operations to a blurry background. The phenomenon of the continuum points at the damage that results from the convenient assumption that complexity can be an afterthought, an add-on delegated to the simple procedure of parametric adjustment in the post-production stage.

Our intuition is that 3D goes through a continuously smooth, multi-dimensional but concentric and loopy flow of assembled technicalities, paradigms and devices that facilitate the circulation of standards and protocols; and hence the constant reproduction of hegemonic metrics for the measurement of volume.[7] Such intuition is nevertheless accompanied by another: that computation can and should operate otherwise. This text therefore makes claims for an attentive praxis that activates a collective technical dissidence from the continuous flows of deadly normality, both in the material sense and in the discursive arrangements that power it.

How is 3D going on?

"Train, Evaluate, Assist." The simulation and training company Heartwood moves smoothly between the classroom and the field to "help operations, maintenance, and field service teams perform complex procedures faster, safer and with less errors." Developing solutions for clients from a wide range of industries (Audi, TetraPak and the United States Secret Service to name a few), Heartwood is proud to insist that it leverages fields as diverse as manufacturing, railroad, utilities, energy, heavy equipment, automotive, aerospace and defense.[8] Their business strategy includes founding principles such as: "There are always new industries to explore — so we do!"[9]

In virtual training solutions like the ones produced by the Heartwood company, we can clearly see how multiple methodical events get arranged in one go. We want to problematize such flows of volumetric techniques and technologies, because of the way this both powers and is powered by the circulation of oppression, exclusion and extraction. The industrial continuum of 3D keeps confirming the deadly normality of European enlightenment, doubtful Judeo-Christian concision, *mono-humanism*,[10] hetero patriarchy and settler colonialism by continuing structures and practices that produce reality. From scientific and metaphysical modes of objectivity into truth, via the establishment of political fictions such as race and gender, to accurate individuality and faithful representation.[11]

The specific vectors that make the Industrial Continuum of 3D indeed continue, are first of all those related to what we call "optimized complexity". It is a particular way to arrange volumetrics in the interest of optimized computation, such as drawing hyper-real surfaces on top of extremely simplified structures or the over-reliance on average simulation. We see this eschewed attention for certain complexities and not for others in how simplified color-coded anatomy travels straight from science books into educational software, and biomedical imaging alike. Divisions between tissues and bones based in standardized category systems organize the relation between demarcated elements in polygonal models, which become hard-coded in constrained sets of volumetric operations and predefined time-space settings, affirmed by scientific nomenclature and recognizable color-schemes that are re-used across software applications. As a result, inter-connective body tissues such as fascia are underrepresented in hyper-real 3D renderings. Thus, the less imperative paradigms that recognize fascia as a key participant in body movement are once again occluded by means of optimization, a very specific industrial phenomenon. As an example of evident continuity by the apparent neutrality of a continuous flow of 3D manners, tissue renderings conserve the way things used to *look like* on 2D anatomy manuals, contributing to the conservation of the way things *are* in terms of anatomical paradigms.

A second vector at work is the *additivist* culture of 3D that thrives on relentless forking and versions to be re-visited and taken back.[12] 3D computation derives agility from the re-use of particle systems, models, data-structures and data-sets to, for example, render grass, model hair or to detect border crossings.[13] Templates, rigs and scenarios are time-consuming to produce from scratch but once their probable topology is set, 3D assets such as "hilly landscape", "turning screw", "first person shooter", "average body"[14] or "fugitive"[15] start to act as a reserve that can be reused endlessly, adjusted and repeated at industrial scale and without ever depleting. Of course that level of flexibility is designed and maintained under positive values such as agility, efficiency and even diversity, but more often than not, their ongoing circulation leads to extreme normalization. With this, we want to point out the fiction of having many options to grab from, which is precisely the settler illusion of the accessibility of resources to take and run with. It still depends on an economy of *asset scarcity*, or even worse: an economy of scarcity that bases its sense of technical abundance on a set of finite, regularized elements.

In addition, volumetrics depends more than other screen based environments on normalized viewing interfaces which makes military training sets and viewing environments for biomedical images follow the exact same representational logic. This is where the techno-scientific paradigms of mandatory projections, perspectives, topology based on binary separations between inside and outside, polygonal treatment, Cartesian axes, Euclidean geometries and so forth are being leveraged to relentlessly spread similar techniques across different corners of practice. Polygonal models travel all too easily between applications because their viewing environments are already standardized. Despite the work of feminist visual culture or cubist avantgardes that have made representation a political issue, perspective devices, anatomy theaters or cartographic projection are once again normalized as cultural standards.[16]

The specific manners in which the techno-sciences historically present metrics of volume nest in distinct fields: from spectacle to control, from laboratories to courts of justice, from syllabi to DIY prototypes, from architecture studies to mining pits. When those manners circulate from one industrial field to another, along vectors that relegate difference and complexity to the background, they reaffirm quite probably the very probable colonial, capitalist, hetero-patriarchal, ableist and positivist topology of contemporary volumetrics. This nauseating and intoxicating setup of variability and rigidity produces the establishment of a universal mono-culture of 3D.

To highlight the continuity of normalizing forces, is our way to critically signal a globalized technocratic behavior based on the accumulation of sameness and repetition, rather than one attuned to the radical, mutating and interconnected specificity of something as wide and multi-modal as the volume of differentiated bodies. 3D models seemingly travel with ease, and this particular easiness facilitates the erasure of politics and the reaffirmation of a central norm. It means the patriarchocolonial linear representation of measurable volumes ends up with providing only with sometimes modular, sometimes fungible entities, circulated by and circulating the everlasting convenience of Modern canons. By Modern convenience, it has become easy to represent distinct elements, but near impossible to engage with inter-connective structures.[17]

Volumetric sedimentation

The monomers can be grouped into segments like Lego pieces to construct functional protein-mimics. "Compare this to how cars are built," said Xu. "There are different models, colors and shapes, but they all contain important parts such as an engine, wheels and energy source. For each part, there can be different options, such as gas or electric engines, but at the end of the day, it's a car, not a train." Xu and her team designed a library of polymers that are statistically similar in sequence, providing newfound flexibility in assembly.[18]

Contemporary biomedical engineering relies on computer generated 3D imagery for inventing materials, pharmaceuticals and fuels and for predicting their behavior. The monomers that Xu and her team compare to a car or a train, are synthetic proteins that were designed using 3D models of cylinders, spirals and spheres.[19] The ease by which a researcher compares a fictional membrane to the car industry is a banal example of how in the hyper-computational environment of biomedical engineering, the interaction between observation, representation, modeling and prediction is settling around — once again — probable patterns.

When the Modern Man finished threading the frame of his latest invention, the perspective device, he could not even start to imagine that centuries later this would be the universally accepted paradigm for representing masses of volume in space.[20] The becoming-paradigmatic of perspective from a static single point has gained terrain through years of artistic, scientific and technical usage throughout realms as diverse as fresco painting or the more recent establishment of a cinematic language. And just as one-point perspective made it all the way from Modernity to our present day, so did other even older paradigmatic techniques such as Cartesian axes, Euclidean geometry, cartographic projection or cubic measurement. These paradigms have been assimilated and naturalized to such an extent that they each lost their own history and have become inseparable from each other, interlocking in ways that have everything to do with the way they support the Modern project. In the current formation, they keep reinforcing each other as the only possible form of representation and thus reality.[21] Their centrality in all found analysis of volume in the world means nothing less than a daily imposition of Euromodern values, modes and techniques of study, observation, description and inscription of the complexity around.[22] In other words: volumetrics are being established due to the multi-vectorial political agenda of Modern technosciences, which is directly entwined with commercial colonialism and Western supremacy.

Despite daily updates, the industrial continuum of 3D is not a changing landscape even if it seems to rely on flow. We can see all sorts of 3D devices and standards circulating in a continuous current from one industry to another, but they persistently move towards a re-establishment of the same, both in terms of shape and of value. Our aim is to understand the paradigms they keep carrying along, and to attend to the assumptions, delegations and canons they impose over matter and semiotics when keeping their business as usual. We suspect there is a rigidification in the establishment of what circulates and what doesn't and we need to see where that persistence hangs from, and how it came to be settled. What are the cultural logics underlying 3D technologies, that turn them into a rigid regime?

One key aspect of the very specific settling of 3D, is that they settle in flow. It is through use and reuse that the establishment of values and manners gets reinforced. A kind of technocratic sedimentation of protocols, standards, tools and formulas which leaves a trace of what is possible in the circuit of volumetrics. The behavior of this sedimentation implies that things just happen again because they happened already before. Every time a tool is adopted from one industry into another, an edge is re-inscribed in the spectrum of what is possible to do with it. And every time the same formula is applied, its axiom gets strengthened. This ongoing settling of the probable in volumetrics comes with its own worlding: it scaffolds the very material-semiotics of what world is to be done, by whom, and by what means. If software making is indeed worldmaking, the settlement of volumetric toolkits and technoscientific paradigms affects what worlds we can world.[23]

For those of us who feel affected by the Cartesian anxiety of always feeling backward[24] in a damaging axiomatic culture of assemblage and measure-all-this-way, it is important to make explicit the moves that reified what it ended up being: an exteriority-less industrial regime based on scientific truths that are being produced by that same regime. It is evident that volume counts a lot in how it came to ostent value, but how does it count and how is it counted? Was it the car industry, that settled values and forms before the Lego blocks appeared? Was it the Lego paradigm of assemblage, that was settled as a reference for biomedical researchers to use it for the predictions in their screens and speeches? The befores and afters matter in this bedrock of shapes and values, as they are telling for what is probably going to happen next.

Over the years, we detected a number of sedimenting behaviors or volumetric probables. The first is *externalizing implications*. The outsourcing of labour and responsibilities is ubiquitous in most industrial computing, but takes a specific shape in the industrial continuum of 3D. Through a strictly hierarchical mode of organization, tasks, roles and all labour-related configurations of relationality persistently, the command is kept in the hands of a privileged minority. Their agendas set industrial priorities but without committing to specific fields or areas of application, therefore avoiding all liability. This adds up to an outsourcing of responsibilities to less powerful agents, such as confronting users with just Yes/No options for agreeing with terms and conditions, or the delegation of energetic costs to the final end of the supply chain.

The need for dealing with computational complexity when rendering volumetrics, leads to an over-reliance on socio-technical standards and protocols that become increasingly hard to undo. *Rigging simplification* refers to the obfuscated reduction inherent in particle systems, for example. A limited set of small samples or 'sprites' is randomized in order to suggest endless complexity. Another example is the way inside and outside is plotted through polygon meshes in CAD files. This technique produces a faster rendering but settles a paradigm of binary separation between interior and exterior worlds. The same goes for the normalized logics of rendering graphics with the help of ray-tracing techniques that demand planar projection for resolving a smooth move between 2D and 3D.[25]

Convenient universalism is how we refer to the way volumetrics technically facilitate modes that avoid dissent, that do not stay with complexity or how all matter becomes equally volumetric before the eyes of the 3D-scanner. Because a virtual dungeon can be rendered with the help of ray-tracing, do the same representational conventions actually apply to dead trees, human brains, aquifers, rhinoceroses and plant-roots? Convenient universalism does not bother to include nuances of minoritarian proposals in mainstream industrial development. It allows ongoing violence to take shape as reasonable, and common sense.

Then, there is the sedimentation of *persistent hyper-realities*. The Continuum operates well when aligning so-called truths, with systems of verification, and performing objectivity. It is not a surprise that it is at ease with Modern scientific and cultural paradigms; its values and assumptions co-construct each other. This is both confirmed and suggested by the over-presence of tools for segmentation and foreground-background separation.[26]

And last but not least, we can speak of *streamlined aesthetics* as a sedimented behavior. It can be confirmed that as the continuum circulates, the aesthetics of tools and their outcomes flatten. The same operations hide behind layers that look the same. Similar procedures are offered by devices that look alike. WYSIWYG interfaces were smoothly adjusted to the machinery of measuring volumes for any purpose... and what sediments in that process is just a sharp similarity all the way along. The aesthetic canon involves equilibrated proportions, hyperrealism and an evident optimization of rendering maneuvers.

The cultural logic of 3D is tied to the ongoing settlement of a legacy of standardization, but also to a history of converging the presences of hugely diversified entities under a rigid regime. This volumetric regime is sustained by vivid Modern techniques, vocabularies, infrastructures and protocols. Or to put it bluntly: the calculation of what it takes to count via the x, y and z axes depends on modes that are far from neutral, and of course are not innocent. The technoscience of volumetrics was settled while being already entangled with a whole world in and of its own.

The Possible Continuums of 3D

In the previous sections we spent some time unpacking how 3D circulates through its industrial continuum and what is sedimented as a result. We clarified what needs to be radically changed or directly abolished to get at a possible volumetrics that can happen non-industrially or at least is less marked by industrial, solutionist values. As we have seen, the industrial continuum of 3D settles and flows in particular ways, making its way through business as usual. It's self-fulfilling moves produce increasingly normed worlds that are continued along the axes of the probable. In this last section, we would like to see what other forms of volumetric continuation, circulation and settlement might be quite possible, as a way to world differently. To find another "how" that can stay with complexity and will not negate, facilitate or altogether erase other modes of existence, we'll need to reorient 3D from a trans*feminist perspective, and move obliquely towards 3D that can go otherwise.

Could an ethics and politics committed to volumetric complexity emerge from reverse-engineering the ebbs and flows of industrial affection? Our first task is to rescue *continuity* from the claws of the established, the normed and the Modern. Against the unbearable persistence of 3D, discontinuity, latency and un-settlement are evident counterforces only as long as they engage with resisting that which 3D settles by flow: neoliberal accumulation, colonial commercial normativity and one-directionality. An affirmative volumetrics does not reject or dismiss the power of volumetrics as a whole, or give up on continuity altogether either. As Donna Haraway asks in conversation with Cathy Wolfe: "How can we truly learn to compose rather than decry or impose?"[27]

We compiled a list of proposals for what we suspect are more affirmative ways, suggestions for dealing with the "volumetric probables" that emerged from our research endeavor so far. They are proposals which are each "nothing short of a radical shift in how we approach matter and form".[28] What is important to keep in mind, is that *none of these are in fact impossible to implement*, so come on!

Remediating Cartesian anxiety: What if we decide to use six instead of four axes, twelve instead of three or zero instead of n? What if we take time to get used to multiple paradigms for orientation, instead of settling for only one regime? Letting go of the finite coordinates of x, y, z and t could be a first step to break with the convenient reductions of parallel and perpendicular assumptions. It's implementations might require rigorous inventions with a transdisciplinary attitude, but we can afford them if what is at stake is to re-orient volumetrics for non-coercive uses, right?

Paranodes to ever-polygonal worlds: By paying attention to the paranodal in ever-polygonal worlds, the simplistic dominance of node-centricity might quickly shift to entirely different topological articulations.[29] This would allow other imaginations of relationality, this time not along the vectors of sameness and similarity but emerging from the undefined materiality of what's there, and what was underrepresented by paradigmatic techno-sciences.

Extra-planar projections: If the distance between 3D and 2D was not to be crossed quickly and straight, but allowed for curves, meanders and loops, then a whole technoscience of dissimilarity and surprise collinearity would emerge. We know the cartographies of complexity are already there, but we just have been lacking the means for their representation, their analysis and their use. Such extra-planar projections would intervene the world with a realm of possibilities in the in-between of 2D and 3D, not assuming the axioms of linear projection but rather convoking the playful articulations of elements diffracted inwards, detailing a scape of situated 2.1D, 2.5.3SD, 2.7Dbis and 2.999999D. The cartographic computation of the possible then becomes a latent one of unsolved folds, abrupt edges, unfinished integers and inaccurate parallels.

Multi-dimensional depth: What background-foreground mergings can we invent for the multidimensional analysis of deep matterings besides volumetrics? Matter is not volume so we need other arrangements of depth and density than the calculating measurings of dimensional worlds. Switching, blurring and blending what comes to the fore with what usually stays behind declutches attention from the binary back-front and inside-outside divides, thickness becomes an area in need of subtle study and nuanced formulations. When the surveillance camera is turned onto the policeman, violence does not go away. But there might be ways to hold paths and crossings in mutual affection and radical sustainability. If capturing would be about soLiDARity instead of policing, about flourishing instead of conservation, about density instead of profiling then fights for social justice might have a chance to reclaim the very dimensions where mundane violence is executed on a daily basis..

Fits-and-starts-volumetrics: Which transformative moves can hold time beyond constant speed, agile advancement and smooth gait? As we learned from Heather Love and her understanding of queer life as constantly feeling backward,[30] as well as from from crip technosciences:[31] linear time is a problematic norm that will always confirm and appreciate what goes forward. In any case, Possible Volumetrics can not be aligned with it. Time as mattered through computation (4D) works too hard on appearing continuous. We propose to use that energy for flowing with what gets crooked and throttled, to move with the flutters and stotterings. Along this text, we tried to show the continuous problematic of the industrialization of 3D, in order to convoke a possible volumetrics that could do 3D otherwise.

In case these proposals feel too hard or even impossible to implement, remember that this sense is always the effect of hegemony! Abolishing the Industrial Continuum of 3D means to place it at the eccentric core of a kind of computing that dares to world without patriarcho-capitalist and colonial structures holding it up.

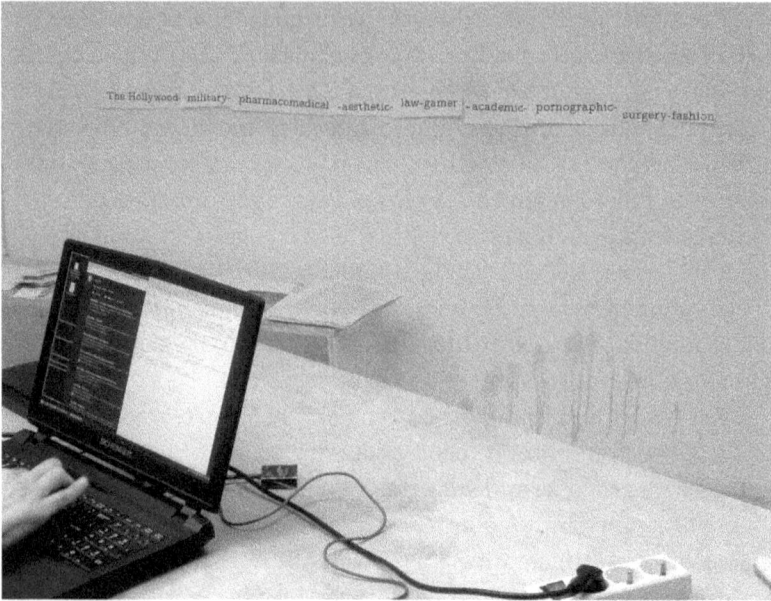

The Industrial Continuum of 3D emerges during "Collective inventorying", Akademie Schloss Solitude, Stuttgart, 2017

"The Industrial Continuum of 3D", fanzine, Barcelona, 2017

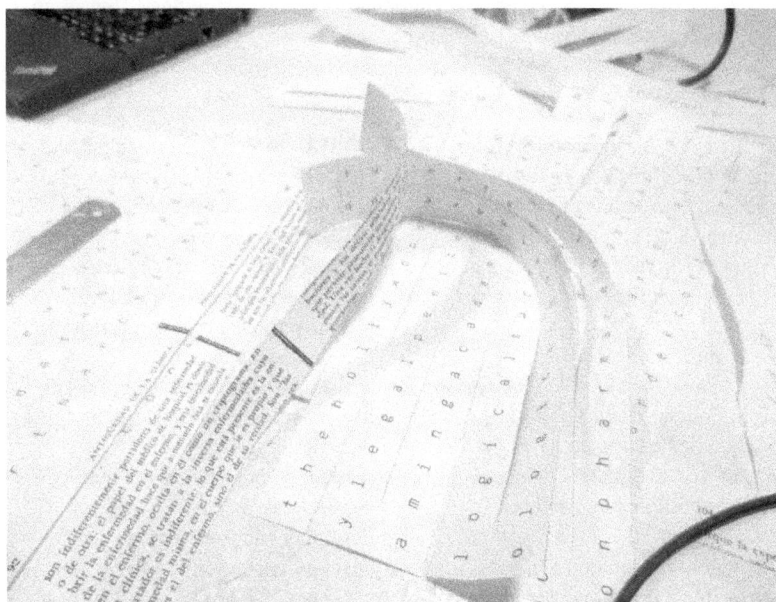

Exploring the continuum with participants in "Imagined Mishearings," Hangar (Barcelona, 2017)

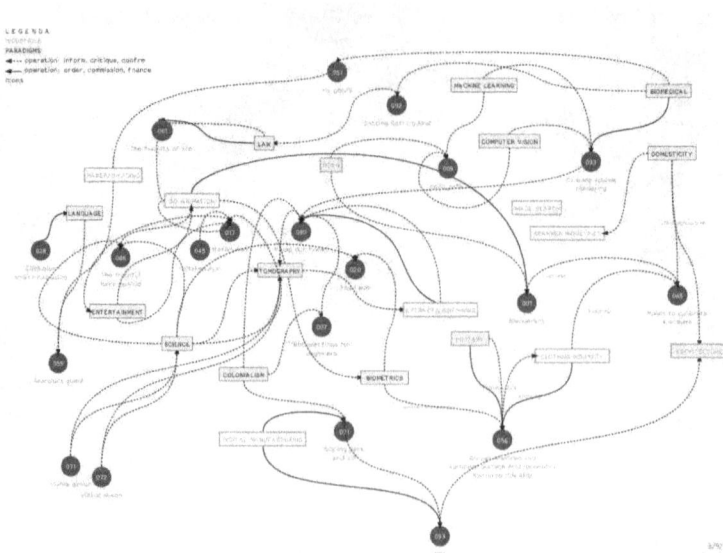

A diagram of The Industrial Continuum of 3D for the workshop "Continuous corpo-realities <-> diagramming probabilities and possibilities!", University of Sussex, Brighton, 2018

Notes

1. ↑ "Item 125: Disappearing around us," *The Possible Bodies Inventory*. Source: Elizabeth Claire Alberts, Mongabay, 21 October 2020, "The rhino in the room: 3D scan brings near-extinct Sumatran species to virtual life".

2. ↑ Loren Britton, and Helen Pritchard, "For CS," *interactions* 27, 4 (July-August 2020), 94–98, https://doi.org/10.1145/3406838.

3. ↑ The British Standard Organisation defines Building Information Modeling (BIM) as: "Use of a shared digital representation of a built asset to facilitate design, construction and operation processes to form a reliable basis for decisions." "BIM - Building Information Modelling - ISO 19650," BSI, https://www.bsigroup.com/en-GB/iso-19650-BIM/.

4. ↑ GAFAM refers to the so-called Big Five tech companies: Google (Alphabet), Amazon, Facebook, Apple, and Microsoft.

5. ↑ "In this way, our contemporary encounters with data extend well beyond notions of design, ease of use, personal suggestion, surveillance or privacy. They take on new meaning if we consider the underlying principles of mathematics as the engine that drives data towards languages of normality and truth prior to any opera-tional discomforts or violences." Ramon Amaro, "Artificial Intelligence: warped, colorful forms and their unclear geometries," in *Schemas of Uncertainty: Soothsayers and Soft AI*, eds. Danae Io and Callum Copley (Amsterdam: PUB/Sandberg Instituut), 69-90.

6. ↑ Helen Pritchard, Jara Rocha, Femke Snelting. "Figurations of Timely Extraction," *Media Theory* 4, no. 2 (2020): 158-188.

7. ↑ "Logistics is straight in that metrically degrading way. This is its murderousness, its refusal to attend to contour, its supervisory neglect and, also, its wastefulness, its continual missing of all in its inveterate grasping of everything." Steffano Harney and Fred Moten, *All Incomplete* (New York: Minor Compositions, 2021), 105.

8. ↑ "Heartwood Simulations & Guides," accessed April 3, 2021, https://hwd3d.com/3d-interactive-training.

9. ↑ "New industries. There are always new industries to explore — so we do! We ask ourselves questions like, 'Will 3D Interactive technology be of interest to the healthcare industry when considering medical device training?' Maybe — but we won't know till we try." Raj Raheja, "When Perfection Is A Little Too Perfect: 3 Ways to Experiment," accessed April 3, 2021.

10. ↑ Katherine McKittrick, *Sylvia Wynter: On Being Human as a Praxis* (Durham: Duke University Press, 2015).

11. ↑ Paul B. Preciado, "Letter from a trans man to the old sexual regime," *Texte zur Kunst*, (2018), https://www.textezurkunst.de/articles/letter-trans-man-old-sexual-regime-paul-b-preciado/.

12. ↑ See for example: "Item 019: The 3D Additivist manifesto," *The Possible Bodies Inventory*, 2015.

13. ↑ See: Jara Rocha, Femke Snelting, "So-called Plants," in this book.

14. ↑ See: Jara Rocha, Femke Snelting, "MakeHuman," in this book.

15. ↑ See: Jara Rocha, Femke Snelting, "So-called Plants," in this book.

16. ↑ Countless thinkers from Svetlana Alpers, to bell hooks, Suzanne Lacy, Peggy Phelan, Elisabeth Grosz and Camera Obscura Collective have critiqued the implicit assumptions in representation. "(R)epresentation produces ruptures and gaps; it fails to reproduce the real exactly. Precisely because of representation's supplemental excess and its failure to be totalizing, close readings of the logic of representation can produce resistance, and possibly, political change." Peggy Phelan, *Unmarked: The Politics of Performance* (London: Routledge, 2003), 3.

17. ↑ See: Jara Rocha, Femke Snelting, "Invasive imaginations and its agential cuts," in this book.

18. ↑ "Item 123: Compare this to how cars are built," The Possible Bodies Inventory. "New discovery makes it easier to design synthetic proteins that rival their natural counterparts," *Berkeley Engineering*, accessed April 3, 2021, https://engineering.berkeley.edu/news/2020/01/new-discovery-makes-it-easier-to-design-synthetic-proteins-that-rival-their-natural-counterparts/.

19. ↑ Protein modeling for prediction: "Model Quality AssessmentPrograms (MQAPs) are also used to discriminate near-native conformations from non-native conformations." "New discovery makes it easier to design synthetic proteins that rival their natural counterparts," *Berkeley Engineering*.

20. ↑ No name needed. Picture an average Modern male, just imagine one that inhabits the very center of power in clear familiarity.

21. ↑ Katherine McKittrick, *Sylvia Wynter: On Being Human as a Praxis*.

22. ↑ Patricia Reed and Lewis R. Gordon define "Euromodernity" in the following way: "By "Euromodernity," I don't mean "European people." The term simply means the constellation of convictions, arguments, policies, and a worldview promoting the idea that the only way legitimately to belong to the present and as a consequence the future is to be or become European." See: Lewis R. Gordon, "Black Aesthetics, Black Value", in *Public Culture* 30, no. 1 (2018): 19-34.

23. ↑ "To provide us with endless a-Modern mestizo, an escape from representational and agential normativities, software CAN and MUST provide the material conditions for wild combinations or un-suspected renders." Jara Rocha, Femke Snelting, "MakeHuman," in this book.

24. ↑ Heather Love, *Feeling Backward* (Cambridge MA: Harvard University Press, 2009).

25. ↑ POV-Ray or Persistence of Vision Raytracer, a popular tool for producing high-quality computer graphics, explains this process as follows: "For every pixel in the final image one or more viewing rays are shot from the camera into the scene to see if it intersects with any of the objects in the scene. These "viewing rays" originate from the viewer (represented by the camera), and pass through the viewing window (representing the pixels of the final image)." "POV-Ray for Unix version 3.7," accessed April 3, 2021, https://www.povray.org/documentation/3.7.0/u1_1.html#u1_1.

26. ↑ See for example the way BIM is used to represent subsurface remnants of demolished structures as separate layers. Gary Morin, "Geospatial World," September 11, 2016, https://www.geospatialworld.net/article/geological-modelling-and-bim-infrastructure/.

27. ↑ Dona Haraway in conversation with Cary Wolfe. Donna J. Haraway, *Manifestly Haraway* (London: University of Minnesota Press, 2016), 289.

28. ↑ Denise Ferreira da Silva. "On Difference Without Separability," in *Incerteza Viva: 32nd Bienal de São Paulo,* ed. Jochen Volz and Júlia Rebouças (Sao Paulo: Ministry of Culture, Bienal and Itaú, 2016), 57-58.

29. ↑ "The instability of paranodal space is what animates the network, and to attempt to render this space invisible is to arrive at less, not more, complete explanations of the network as a social reality." Ulises Ali Mejias, *Off the Network: Disrupting the Digital World* (Minneapolis: University of Minnesota Press, 2013), 153; and Zach Blas, "Contra-Internet," *e-flux Journal* #74 (June 2016).

30. ↑ Heather Love, *Feeling Backward.*

31. ↑ Aimi Hamraie and Kelly Fritsch, "Crip Technoscience Manifesto," *Catalyst* 5, no. 1 (2019).

Signs of Clandestine Disorder in the Uniformed and Coded Crowds

Possible Bodies (Jara Rocha, Femke Snelting)

What are the implications of understanding bodies as political fictions in a technical sense? With what techniques, technologies, protocols and/or technoscientific paradigms are contemporary volumetric forms entangled? How are the probabilities of these technosciences strained by the urgency to broaden the spectrum of semiotic-material conditions of possibility for the bodies present? What worldly consequences does the paradigm of the quantified self bring? Bodies (their presence, their permanence, their credibility, their potential) are affected by the way they are measured, remeasured and mismeasured. This workshop-script was developed for a workshop on speculative somatic measuring and data interpretation. It invites participants to invent other systems of measuring bodies by mixing already existing disciplines or crossings with what is yet to come: anatomy, physics, chemistry, geometry, biology, economics, anthropometry...

Duration: 2 ca. hours; between 6 and 60 participants.

Materials to prepare
- Sheets with *situation*, 1 for each group
- A sheet with an empty legend and space for description, 1 for each group
- A set of Measurement units, printed on small pieces of white paper, 5 for each group: kg (weight), grams (weight), milligrams (weight), tons (weight), ρ (mass per unit, density), red (RGB), green (RGB), blue (RGB), mm (height), cm (height), km (height), mm (width), cm (width), km (width), years (age), mm (diameter), cm (diameter), meters (diameter), cm (radius), m (radius), cm2 (surface area), km2 (surface area), m2 (surface area), (number of corners), (number of limbs), liters (volume), cm3 (volume), BMI (Body Mass Index), likes, IQ...
- Small pieces of colored paper, 2 for each group
- Empty pieces of white paper, 5 for each group
- A set of numbers (12, 657.68787, 24, 345, 0.00012, 2000, 1567, 4...), printed on small pieces of white paper, 5 for each group

Introduction: From the probable to the possible
(10 minutes)

The workshop is introduced by reminding participants of how we as quantified selves are swimming in a sea of data. Bodies (their presence, their permanence, their credibility, their potential) are affected by how they are measured, remeasured and mismeasured. These measurements mix and match measurement systems from: anatomy, physics, chemistry, geometry, biological, economic, biometrics...

Numbers
(10 minutes)

Divide the participants in groups of between three and five participants. Each group selects 5 numbers. Ask if participants are happy with their numbers.

Measurements
(15 minutes)

Remind participants that these are raw numbers, not connected to a measuring unit. Brainstorm: What measurement units do we know? Try to extend to different dimensions, materials, disciplines. Each group receives 5 papers with measurement units.

Bodyparts
(5 minutes)

Groups have received numbers + measurement units. But what are they measuring? Each group proposes 2 body parts and writes them on the colored paper. These can be internal, external, small, composed... Gather all colored papers, mix and redistribute; each group receives 2.

Situation
(5 minutes)

Alphonso Lingis, *Dangerous Emotions* (University of California Press, 2000): "We walk the streets among hundreds of people whose patterns of lips, breasts, and genital organs we divine; they seem to us equivalent and interchangeable. Then something snares our attention: a dimple speckled with freckles on the cheek of a woman; a steel choker around the throat of a man in a business suit; a gold ring in the punctured nipple on the hard chest of a deliveryman; a big raw fist in the delicate hand of a schoolgirl; a live python coiled about the neck of a lean, lanky adolescent with coal-black skin. Signs of Clandestine Disorder in the Uniformed and Coded Crowds."

Drawing and annotating
(30 minutes)
Fill out the legend with the data you received, and draw the so-called body/bodies that appear(s) in this situation. Make sure all participants in the group contribute to the drawing. Circulate or draw together. Fold back the legend and re-distribute the drawings.

Interpretation
(30 minutes)
Each group makes a technical description of the drawing they received and details the measurements where necessary. Possible modes of interpretation: engineer, anthropologist, biologist, science fiction writer...

Reading
(15 minutes)
Re-distribute the drawings and descriptions among groups. Look at the drawing together. Read the interpretations aloud.

So-called Plants

Jara Rocha, Femke Snelting

Spray installations enhanced with fruit recognition applications, targeting toxic load away from precious mangoes;[1] software tools for virtual landscape design economizing the distribution of wet and dry surfaces;[2] algorithmic vegetation modeling in gaming which renders lush vegetation on the fly;[3] irrigation planning by agro-engineering agencies, diminishing water supplies to the absolute minimum;[4] micro-CT renderings of root development in scientific laboratories:[5] all of these protocols and paradigms utilize high-end volumetric computation. Vegetation data processing techniques make up a natureculture continuum that increasingly defines the industrial topology applied to the existence of so-called plants. These techniques integrate 3D-scanning, -modeling, -tracking and -printing into optimized systems for dealing with "plants" as volume.

Thinking with the agency of cultural artifacts that capture and co-compose 3D polygon, point-cloud and other techniques for volumetric calculation, we brought together over a hundred items in The Possible Bodies Inventory.[6] For this chapter, we selected several "computational ecologies of practice" that allow us to "feel the borders" of how so-called plants are being made present.[7] We write "so-called plants" because we want to problematize the limitations of the ontological figure of "plant", and the isolation it implies. This is a way to question the various methods that biology, computer science, 3D-modeling or border management put to work to create finite, specified and discrete entities which represent the characteristics of whole species, erasing the nuances of very particular beings. We are wondering about the way in which computational renderings of so-called plants reconfirm the figure-background reversals which Andrea Ballestero discusses in her study of the socio-environmental behavior of aquifers. This flipping between figures and their ground not only happens because of the default computational gestures of separation and segmentation, but also through cycles of flourishing, growing, pollinating, and nurturing of "plants" that appear animated while being technically suspended in time. Such reversals and fixities are the result of a naturalization process that managed to determine "plants" as clearly demarcated individuals or entities, arranged into landscapes along which their modes of existence develop under predictable and therefore controllable conditions. It is this production-oriented mode that 3D volumetrics seem to reproduce.

The Possible Bodies Inventory is itself undeniably part of a persistently colonial and productivist practice. The culture of the inventory is rooted in the material origins of mercantilism and deeply intertwined with the contemporary data-base-based cosmology[8] of techno-colonialist turbo-capitalism.[9] Inventorying is about a logi(sti)cs of continuous updates and keeping items available, potentially going beyond pre-designed ways of doing and being as proposed by the mono-cultures of what we refer to as "totalitarian innovation", and what Donna Haraway calls "informatics of domination".[10] Inventories operate in line with other Modern devices for numbering, modeling and calculating so-called plants: herbaria, which function as a physical re-collection of concrete plant specimens; genetic notebooks, which trace lines around and between individuals; Latin nomenclatures, which produce and reproduce taxonomies of species within so-called families within so-called kingdoms; sketchbooks filled with naturalist drawings captured during explorations; and even botanic gardens that arrange lively exotic samples to be experienced in the overseas environments of the metropolis. An inventory can be understood as a workspace arranged for constant managerial return, and — in contrast with a collection or an archive — they allow easy access to items for re-ordering, removal or replacement. Just like almanacs used in observatories or taxonomies at museums, inventories play a role in the becoming of computational herbaria as contemporary apparatuses for the production of knowledge, capital and order.

Possible Bodies attends obliquely to the power relations embedded within inventories, because it provides a possibility to open up methods for disobedient action-research. Following trans∗feminist techno-sciences driven by intersectional curiosity and politics, the inventory attempts to unfold the possibilities of this Modern apparatus for probable designation and occupation. Disobedient action-research implies radical un-calibration from concrete types of knowledge and proposes a playful, unorthodox and "inventive" inhabiting of many disciplines, of learning, unlearning and relearning on the go. It also plots ways to actively intervene on the field of study and interlocutes with its communities of concern and their praxis of care. In this chapter, we try to relate to the cracks in the supposedly seamless apparatus of 3D. Curious about the post-exotic[11] rearrangement of methods, techniques and processes that follow the industrial continuum of 3D,[12] we selected various inventory items of vegetal volumetrics to consider the promising misuses of Modern apparatuses, and technical counter-politics as an active matter of care.[13] This text tries to provide with a trans∗feminist mode of understanding and engaging with so-called plants not as individual units, but as vegetal forms of computationally implicated existence.

Vegetal volumetrics

The following two items pay attention to processes of vigilant standardization as a result of collapsing the one with the many. They each apply a disobedient volumetrics to resist naturalizing representation as evidence of a universal truth. The items want to cultivate the ability for response-ability[14] within computational presentations of the vegetal. Instead of the probable confirmation of hyperproductive 3D-computation, these items root for a widening of the possible and other computational ways of rendering, modeling, tracking and capturing so-called plants.

Item 102: Grassroot rotation

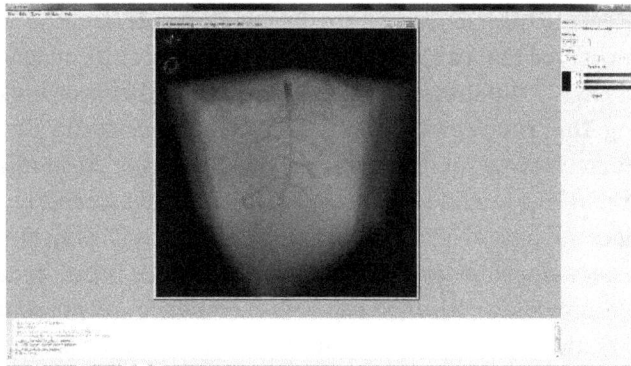

Author(s) of the item: **Stefan Mairhofer**
Entry date: **2 July 2018**
Cluster(s) the item belongs to: **Segmentation**
Image: Segmentation of a tomato root from clay loam using RooTrak[15]

En nuestros jardines se preparan bosques ("In Our Gardens, Forests Are Being Prepared") is a thick para-academic publication on political potential by Rafael Sánchez-Mateos Paniagua, alluding to the force of potentiality that is specific to vegetal surfaces, entities and co-habituating species, which turns them into powerful carriers of political value.[16] Other than productive and extractive, they are informative of the inner functionings, inter-dependencies and convivial delicacies with so-called plants.

Item 102: Grassroot rotation is a poetic rendering of demo-videos that accompany a manual for RooTrak, a software-suite for the automated recovery of three-dimensional plant root architecture from X-Ray micro-computed tomography images. The images we see rotating before us, are the result of a layered process of manual and digital production, starting with separating a grass "plant" from it's connected, rhizomatic neighbors and in that sense, it is a computationally gardened object. The "plant" is grown in a small, cylindrical container filled with extracted soil before being placed in a micro-CT installation and exposed to X-rays. The resulting data is then calibrated and rendered as a 3D image, where sophisticated software processes are used to demarcate the border between soil and "plant", coloring those vessels that count as root in blood red. The soil fades out in the background.

In collaboration with RooTrak, the software package responsible for these images, X-ray microcomputed tomography (μCT) promise access to the living structure through "a nondestructive imaging technique that can visualize the internal structure of opaque objects."[17] But these quantified roots are neither growing nor changing. They rotate endlessly in a loop of frozen or virtual time, which can be counted and at the same time not. The virtualized roots pass through time while the computed loop goes on smoothly... but time does not pass at all. The roots are animated as if lively, but simultaneously stopped in time. Speed and direction are kept constant and stable, providing with an illusion of permanence and durability that directly links this re-presentational practice to the presentational practice of cabinets, jars and frames. The use of animation has been persistent in the scientific study of life, as a pragmatic take on "giving life" or technically re-animate life-forms. After first having claimed the ability to own and reproduce life by determining what differentiates it from non-life, all of this is done in an efficient manner, combining positivist science with the optimization mandate of the industrialized world. But how does the 3D animation complex apparatus do the trick of determining life and non-life? While RooTrak prefers to contrast its particular combination of CT-imaging and 3D-rendering with *invasive* techniques such as root-washing or growing roots in transparent agar, to us this grassroot rotation seems closer to the practice of fixing, embalming and displaying species in formaldehyde.

The tension between animism and animation can be studied from the dimension of time and its specific technocultural maneuvers present in *Item 102*. It helps us see how computed representations of the animated vegetable kingdom continue to contribute to the establishment of hierarchies in living matter. What are the consequences of using techniques that isolate entities, which need complex networks for their basic existence? What is kept untold if different temporalities are collapsed into smooth representations of specimens, as if all happened simultaneously?

Item 033: This obscure side of sweetness is waiting to blossom

Author(s) of the item: **Pascale Barret**
Year: **2017**
Entry date: **March 2017**
Image: 3D print by Pascale Barret[18]

Item 033 features a work by Brussels-based artist Pascale Barret.[19] A 3D object is printed from a volumetric scan of a flowering bush with an amateur optical scanner. The object has nothing and everything to do with so-called plants, because the low-res camera never went through a machinic training process to distinguish or separate leaves.

The software processing the data-points then algorithmically renders the vegetation with an *invented* outside membrane, a kind of outer petal or connective tissue that is sneaked into the modeling stage and finally materialized by the printing device. This invention might look hallucinatory to the eyes of a trained botanist, but for us it is a reminder of the need to re-attune digital tools in a non-anthropocentric manner. Pascale printed the volumetric file at the maximum scale of the 3D printer she had available, breaking the promise of the 1:1 relationship between scanned object and its representation. Because she did not remove the scaffolding that upheld the soft plastic threads during the printing process, these now "useless" elements flourish as twigs once the object had solidified. The item talks to us about a complex switching of agencies. In the first place, the vegetal groupings in *This obscure side of sweetness is waiting to blossom* are formed by the surfaces that the algorithm computes in-between leaves caught by the scanner, trying to connect their wild in-betweens. The leaves defy linear, isolating and rigid capture by taking on the too obedient mathematics that tries to encapsulate each gap and jump. Their surfaces and their positionality invite the computation of a continuous topological surface, based on straight mathematical axioms and postulates. What interests us here is that the axioms nested in the operations of 3D optics and scanning, are stretched towards a *beyond-realistic* materialization. This switching of agencies is operating according to a logic that simultaneously defies the *realistic* establishment of a topological space, while creating a *manifold* that looks Euclidean. "Plants" become computable, accountable, nameable, determined, and discrete without giving in to the promise of mimesis.[20]

In the way "plants" have been historically described, there is an ongoing attempt to fix the zones where they actually can be, become and belong. But looking closely, we can easily identify paranodal spaces in-between the vegetal and other forms of existence, gaps or porous membranes which exist beyond the positive space of nodes and links. These can be seen as void and sterile spaces in-between known entities, but they can also be taken as wide open, inhabitable areas; places to be in-relation that are non-neutral and also not innocent at all: connecting surfaces that provide with the blurring travel from one isolated unit of life onto another, in specific ways. Holes, gaps or even chasms are zones of the world in and for themselves.[21] Mel Y. Chen's work on interporousness tries to come to terms with the way interspecies interabsorbence is prefigured by more-than-human power relations. "The couch and I are interabsorbent, interporous, and not only because the couch is made of mammalian skin. These are intimacies that are often ephemeral, and they are lively; and I wonder whether or how much they are really made of habit."[22] Their work shows how the attempt to separate, segment, identify and onto-epistemologically demarcates sharp edges between the mammalian, vegetal and human modes of existence. It must be considered as a damage due to the persistent cutting apart of dense and complex relational worlds that as a result, do not show cracks and paranodal spaces as inhabitable anymore. How those damaging representations infuse the contemporary computational take on "plants", is a direct consequence of Modern technosciences and their utilitarian, exploitative foundations, based on the fungibility of some matters and the extraction of others. But if we think of seeds blown by the wind, roots merged with minerals or branches grabbing the whole world around them... formerly disposable cracks and gaps also have lively potential for ongoingness, as areas for circulating matters. From useless to blossoming, from separating borders to articulated and activated cracks, we need a persistent flipping of agencies, ongoing "circluding" moves that are difficult, but not impossible to uphold in computed spaces.[23]

Systemic vegetation

In her work on the involution of plants and people, Natasha Myers invites us to consider renaming the *Anthropocene* into *Planthro-poscene* as it "offers a way to story the ongoing, improvised, experimental encounters that take shape when beings as different as plants and people involve themselves in one another's lives."[24] With her proposition in mind, we now move upwards and sideways from the topological attention to surfaces of vegetal specimens, and the way they are cut together and apart by naturalized modes of (re)presentation, to the quantification and tracking of wide and thick surfaces. From Planthroposcene to Plantationcene, this section pays attention to a set of volumetric operations for predicting, optimizing and scaling full areas arranged as gardens, forests, landscapes or plantations in which so-called plants are made part of a system of intensive worlding, not free from similar options of measurement, control and scrutiny.[25]

Item 117: FOLDOUT

Year: **2018-2022**
Author(s) of the item: **HORIZON 2020**
Entry date: **15 July 2020**
Image: "Terrestial scenario (Ground dataset), classification based on radiance: (a) Raw hyperspectral input frame. (b) Corresponding RGB frame. (c,d) Classification without and with spectral overlap suppression, respectively. White: Classified as positive sample. Black: Classified as negative sample."[26]

Item 117 references FOLDOUT, a five year collaboration between various research departments across Europe on border control in forest areas. FOLDOUT aims to "develop, test and demonstrate a solution to locate people and vehicles under foliage over large areas."[27] Dense vegetation at the outer borders of the EU is perceived as a "detection barrier" in need to be crossed by surveillance technology. The project received 8,199,387.75 € funding through the European Union's Horizon 2020 scheme and its central approach is to integrate short- (ground based), medium- (drones), long- (airplane) and very long-range (satellite) sensor techniques to track "obscure targets" that are committing "foliage penetration". FOLDOUT says to integrate information captured by Synthetic-Aperture Radar (SAR), Radio Detection and Ranging (RADAR), Laser imaging, Detection, and Ranging (LiDAR) with Low Earth Orbit satellites (LEO) into command, control and planning tools that would ensure an effective and efficient EU border management.[28]

To detect "foliage penetration", FOLDOUT relies among others on "foliage detection", a technique now also widely used for crop optimization. In agricultural yield estimation or the precision application of pesticides for example, hyperspectral imaging and machine learning techniques are combined to localize leaves and tell them apart from similar shapes such as green apples or grapes. Hyperspectral imaging scans for spectral signatures of specific materials, assuming that any given object should have a unique spectral signature in at least a few of the many bands that are scanned. It is an area of intense research as it is being used for the detection, tracking and telling apart of vehicles, land mines, wires, fruit, gold, pipes and people.[29]

FOLDOUT is a telling example of the way "fortress Europe" shifts humongous amounts of capital towards the entanglement of tech companies with scientific research, in order to develop the shared capacity to detect obscurity at its woody barriers.[30] By sophisticating techniques for optimized exclusion, negation and expulsion, Europe invests in upgrading the racist colonial attitude of murderous nation states. How to distinguish one obscureness from another seems a banal issue, seen from the perspective of contemporary computation, but it is deeply damaging in the way it allows for the implementation of remote sensing techniques at various distances, gradually depleting the world of all possibility for engagement, interporousness and lively potential. In the automation of separation (of flesh from trunk, of hair from leaves, of fugitive from a windshaken tree) we can detect a straightforward systematization of institutional violence.

Apples are red,
leaves are green,
branches are brown,
sky is blue
and the ground is yellow.

Almonds are blue,
leaves are red,
branches are black,
sky is blue
and the ground is white.

Apples are red,
leaves are green,
branches are brown,
sky is blue
and the ground is yellow.

Mangoes are black,
leaves are white,
branches are yellow,
sky is red
and the ground is white.

Mangoes are red,
leaves are blue,
branches are green,
sky is black
and the ground is yellow.

Fugitives are blue,
branches are red,
sky is yellow,
leaves are black
and the ground is white.[31]

Item 118: Agrarian Units and Topological Zoning

Entry date: **15 July 2020**
Cluster(s) the item belongs to: **Segmentation**
Inventor(s) for this item: **Abelardo Gil-Fournier**
Image: Agribotix™ FarmLens™ Image Processing and Analytics
Solution, viewed on WinField's Answer Tech® Portal[32]

Item number 118 features the research and practice of Abelardo Gil-Fournier, and with him we learn how agriculture is volumetric. He quotes Geoffrey Winthrop-Young to highlight how elemental "agriculture [...] is initially not a matter of sowing and reaping, planting and harvesting, but of mapping and zoning, of determining a piece of arable land to be cordoned off by a boundary that will give rise to the distinction between the cultivated land and its natural other".[33] Gil-Fournier continues:

> *However, this initial two-dimensional demarcation gives rise to a practice that can be further understood when the many vertical layers that exist simultaneously above and below the ground start to be considered. From the interaction of synthetic nutrients in the soil with the roots of the plants, to the influence of weather or the effect of both human and machinic labor, agriculture appears as a volumetric activity.*[34]

The inclusion of such massive vertical management of soil with the aim of fertilizing it, reorients agriculture from an engagement with surface to the affections of scaling up-and-down the field. To explain the way soil matter is turned into a "legible domain", Gil-Fournier takes as a case study the Spanish "inner colonization" that organized land and landscapes for plantation and irrigation. Through those studies, it is made materially explicit how the irrigation zones configure a network-like shape of polygonal meshes that distribute and systematize the territory for a sophisticated exploitation of its vegetal potentials. In Francoist Spain, under a totalitarian regime of autocracy, inner colonization was the infrastructural bet which provided the nationalist project with all needed resources from within, as well as with a confident step into the developmentist culture of wider Western, Modern economies. Gil-Fournier's work facilitates a departure point for a study of the legacies carried by contemporary hyper-computational applications that are currently being tested to, for example, analyze the seasonal evolution of gigantic agro-operations or to detect the speed by which desertification reveals the diminishing of so-called green areas.[35]

Recent space imaging developments have given rise to a spread of commercial services based on the temporal dimensions of satellite imagery. Marketed under umbrella terms such as environmental intelligence, real-time Earth observation or orbital insight, these imaging projects deliver the surface of the planet as an image flow encoded into video streams, where change and variation become a commodified resource on the one hand, as well as a visual spectacle on the other."[36]

The structural connection between volumetrics and soil observation unfolds when soil itself is treated as a segmentable and computable surface for purposes as different as climate change monitoring, new resource location or crop growth analysis and maintenance. The big-scale top-to-bottom agro-optimization of vegetal surfaces by hyperproductive means, urges us to consider the Plantationcene:

Plantation as a transformational moment in human and natural history on a global scale that is at the same time attentive to structures of power embedded in imperial and capitalist formations, the erasure of certain forms of life and relationships in such formations, and the enduring layers of history and legacies of plantation capitalism that persist, manifested in acts of racialized violence, growing land alienation, and accelerated species loss.[37]

The "enduring layers of history and legacies of plantation capitalism" can be read in the hyper-quantified vegetation praxis that we observed, a continuous flow of similar logics and logistics that forms what we elsewhere termed "The Industrial Continuum of 3D".[38]

Scaling up and moving sideways, the Plantationcene joins the Planthroposcene to tell stories of how the systematization of vegetation happens partially through a set of volumetric operations, for the sake of vegetal extraction and intensive multi-planar exploitation. Such ongoing surveillance of growth continues to produce and reproduce systemic oppressions, and asks us to stay attentive to and eventually twist the flattening monocultures that 3D tools and devices engender.

Lively math

In the first two sections, we discussed the paradigm of "capturing" by scanning plants, and the politics of vegetal topology. Now we would like to turn to the particular technocultural conflation of "beauty", "scientific accuracy" and "purpose" that is intensified in the modeling of 3D vegetals. We insist that this type of conflation is cultural, because it explicitly depends on a classic canon that turns only certain equilibriums and techniques into paradigmatic ones. This section tries to get a handle on the many levels of aesthetic and semiotic manipulation going on in the "push and pull" between botany and computation. It is written from an uncalibrated resistance to the violence inherent in this alliance, and the probable constraints that computation inflicts on the vegetal and vice versa.

Item 119: IvyGen

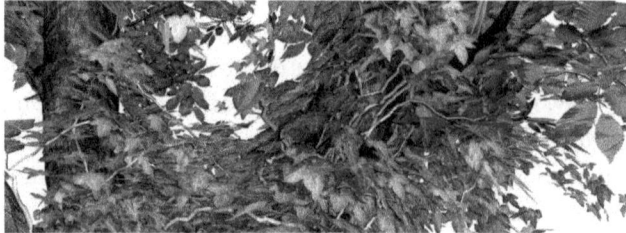

Author(s) of the item: **Thomas Luft**
Year: **2008**
Entry date: **18 September 2020**
Image: IvyGen, screenshot[39]

Item 119 is called IvyGen, after a small software tool developed in 2007 by a now retired computer graphics professor, Thomas Luft. Luft was looking for a "sample scene" for his work on digitally emulated watercolor renderings: "I was thinking of something complex, filled with vegetation — like trees overgrown with ivy. Fortunately I was able to implement a procedural system so that the ivy would grow by itself.

The result is a small tool allowing a virtual ivy to grow in your 3d world." 10 years later, we find Luft's rudimentary code back as the *Ivy Generator add-on* which can be installed into Blender, a free and open-source 3D computer graphics software suite. The manual for IvyGen add-on reads as follows:

1. Select the object you want to grow ivy on.
2. Enter Edit Mode and select a vertex that you want the ivy to spawn from.
3. Snap the cursor to the selected vertex.
4. Enter Object Mode and with the object selected: Sidebar ▸ Create ▸ Ivy Generator panel adjust settings and choose *Add New Ivy*.[40]

The smooth blending of computational affordances with natural likeness that was already present in Luft's original statement (promising "ivy that would grow by itself" in "your 3d world")[41] is further naturalized in these simplified instructions. The slippage might possibly seem banal because computational vocabulary already naturalizes vegetal terms such as tree, root, branching, seeds and so on to such an extent that the phrase "Select the object you want to grow ivy on" at first causes no alarm. It is common in modeling environments to blend descriptions of so-called bodies with those of their fleshy counterparts. This normalized dysphoria is considered a short-cut without harm, a blurring of worlds that does not signal any real confusion or doubt of what belongs to what. The use of "plant" when "so-called plant" would be more accurate, effectuates a double-sided holding in place, that ignores the worlding power of modeling so-called ivy in computation, and removes the possibility for these ivies to make a difference.

Non-computational ivy is a clear example of *symbiogenesis*,[42] meaning that it is materially, structurally and behaviorally always-already implicated in co-dependence with other structures, vegetal or not, straight or crooked, queer or dead. But the vegetal modeling in IvyGen takes another route. So-called plants are drawn from one single starting point that then is modulated according to different computed forces. Parameters allow users to modulate its primary direction of expansion (the weighted average of previous expansion directions), add a random influence, simulate an adhesion force towards other objects, add an up-vector imitating the phototropy of so-called plants, and finally simulate gravity. The desire and confidence by which this procedural system makes Ivy "grow" itself, is not innocent. Technically, Ivy Gen implements a Fibonacci sequence complexified by external forces that act as "deviators", and variation is the result of a numerical randomization applied after-the-fact. The Fibonacci sequence is a string of numbers that describes a spiral that mathematician Fibonacci coined as "golden proportions", and similar ratios can allegedly be found in biological settings such as: tree branching, the fruit sprouts of a pineapple, the flowering of an artichoke, an uncurling fern, and the arrangement of a pine cone's bracts. It became a pet project for nature lovers, math enthusiasts and 3D-modellers who create an ongoing stream of more or less convincing computer programs and visualizations celebrating algorithmic botany or computational phyllotaxy, the botanical science of leaf arrangements on plant stems. The Fibonacci sequence is a mathematical construct that has just the right combination of scientific street cred, spiritual promise and eloquent number wizardry to convincingly bring patterns in "nature" in direct relation to math and computation, confirming over and over again that aesthetics and symmetry are synonymous and that simple rules can have complex consequences. The obsession with computed leaf patterns reinforces the idea that dynamic systems are beautiful and predictable. In turn, these programs confirm how spiraling "plant" patterns are not just elegant, but that they are inevitable. They can be decoded like computer software, and in the process, computation becomes as stunning as nature itself.[43]

Like in many other modeling set-ups for simulating biological life, IvyGen aligns 3D computation with phyllotaxy without any reservations. It constructs so-called plants as autonomous individuals through "expansion patterns", which are straight at the core. This is not surprising because the procedural conditionings of computation seem to make political fictions of life which provoke technocratic and scientific truths of so-called bodies, more easy to implement than others.[44] IvyGen re-asserts a non-symbiogenetic understanding of evolution and ecology where growth is a deformation of the symmetrical, a deviation after the fact. Queer angles can only arrive afterwards, and are always figured as disruption, however benign and supposedly in the interest of convincing realism. Luft clarifies that "the goal was not to provide a biological simulation of growing ivy but rather a simple approach to producing complex and convincing vegetation that adapts to an existing scene".[45] The apparent modesty of the statement confirms that even if the goal has not been to simulate non-computational ivy, the procedural system is seen as a "simplified" approach to actual biological growth patterns, rather than an approach that conceptually and politically differs from it. The point is not to correct IvyGen to apply other procedures, but to signal that the lack of problematization around that rote normalization is deeply problematic in and of itself.

Item 120: Simulated dendrochronology for demographics?

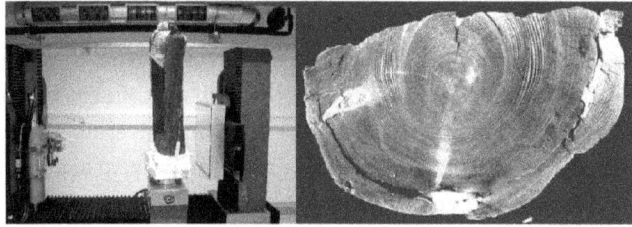

Author(s) of the item: **Pedro Cruz**, **John Wihbey**, **Avni Ghael**, **Felipe Shibuya**
Year: **2017-2018**
Entry date: **18 September 2020**
Image: Tailoring X-ray imaging techniques for dendrochronology of large wooden objects[46]

Dendrochronologist study climate and atmospheric conditions during different periods based on tree-ring growth in wood.[47] This particular scientific way to relate to life has to be individual-centered in order to make trees emerge in their ideal form. Dendrochronology is based on seeing a tree as a perfect circle, assigned to such individual. All variations along that specimen's existence are just the result of modifications radiating outwards from the perfect mathematical zero point. Instead of departing from a complex environment full of forces interlaced in the midst of which a tree grows, dendrochronology reads the aberrations and deviations from the geometrical circle as exceptional interventions deforming its concentric expansion, and by doing so time and time again re-confirms and projects the idealized geometry as the desired centered and balanced life-pattern for a tree. This approach confirms the understanding of the plant's growth as a predictable phenomenon (i.e. beautiful), which make it become a vector into the probable (i.e. extractive, exploitative ideology) and distances it from the surprise ontologies of the possible.

The project *Simulated Dendrochronology of U.S. Immigration*[48] takes dendrochronology as a visual reference to represent the development of US demographics by immigration as "natural growth". "The

United States can be envisioned as a tree, with shapes and growing patterns influenced by immigration. The nation, the tree, is hundreds of years old, and its cells are made out of immigrants. As time passes, the cells are deposited in decennial rings that capture waves of immigration." The work by Pedro Cruz, John Wihbey, Avni Ghael and Felipe Shibuya two Kantar won a prestigious design prize[49] and seems to be generally read as a benevolent rendering of immigration as being "in the nature" of the United States. As some of their colleagues note, they make "immigration look positive, natural, and beneficial".[50] But by visualizing immigration data as a severed tree, the infographic almost literally flattens the lively complexity of demographics, first of all by essentializing the category of the nation state as formative of population evolution. Second, by offering only an accountability of "entrances" and not "exits" (e.g. not accounting for deportations). And last but not least, by imposing a mechanism of naturalization over a social behavior inextricably linked to economic, cultural and political conditionings.

As an invasive volumetric study that studies growth with the help of cylindrical samples after very precise planar drilling, dendrochronology as a technique also carries the story of how Modern technosciences in one way or another gaslight the borderline between existence and representation. In other words: the horizontal strata of tree rings present a specific complex and rich worlding, while the disciplinary study of them overimposes a view of what ought to be through the application of comparative and quantitative methods that foreground average behavior as well as the measuring of the distance of that specimen from an ideal representation of its species. How could dendrochronology inform on difference, instead of imposing ideals, inviting the probable and avoiding forgiving comparisons of nation-state demographics as if they were "resembling a living organism", only subjected to climate inclemencies? The worrying benevolence in the data visualization work, trying to naturalize immigration via the greenwashing figuration of a tree trunk cut, alerts us to technocultural leaps. The equation of vegetal symmetry, straightness and proportionality has deep implicancies. The aesthetic and semiotic manipulation which benevolent data visualization accomplishes, removes responsibility for the conditions that produce its necessity. The naturalization of the thick damage of migrating experiences, violently revalidates a world made up of borders and states that kill. We simply can not afford more deadly simplifications.

Cracks and flourishings

In a conversation with Arjuna Neuman, Denise Fereira da Silva contrasts her use of the term "Deep Implicancy" with that of "entanglement": "The concept of Deep Implicancy is an attempt to move away from how separation informs the notion of entanglement. Quantum physicists have chosen the term entanglement precisely because their starting point is particles (that is, bodies), which are by definition separate in space."[51] She insists that by paying attention to the relations between particles, their singularity as entities (just as so-called plants, leaves or petals) is being reconfirmed. In the very matter of the notion, implicancy or "implicatedness" can be understood as a circluding[52] operation to the notion of entanglement, in the sense that it affirms a mutual constitution from scratch.[53]

When attempting a disobedient action-research in volumetrics oriented towards so-called plants, we try to start from such mutuality to understand at least two things. First, what are the cracks in the apparatus of contemporary 3D that is too-often presented as seamless. How and where can those cracks be found and signaled, named and made traceable? Second, how can we provoke and experience a flourishing of volumetric computation otherwise, attentive to its implicancies and its potential to widen the possible? In *Vegetal volumetrics*, those surfaces that provide bridges for jumping from one unit of life to another, are made tangible in *Item 033: This obscure side of sweetness is waiting to blossom. Item 102: Grassroot rotation* exposed the consequences of contrasting life and non-life all too graphically. These items call for different a-normative interfaces; ones whose settings would not already assume the usefulness or liveliness of one area over the uselessness and backgroundness of another. *Systemic vegetation* brought two items together to ask how plants are made complicit with deadly operations. *Item 117: FOLDOUT* points at the urgency to resist the automation of separation as a way to block the systematization of institutional violence. *Item 118: Agrarian Units and Topological Zoning* showed how staying with the volumetric traces, keeping memories of and paying attention to certain forms of life and the relationships between such formations might open up possibilities for coming to terms with the systemic alienation going on in plantations. The last section, *Lively math*, investigated the stifling mutual confirmation of math and so-called plants as "beautiful", "inevitable" and "true". *Item 119: IvyGen* proposes non-normative dysphoria to queer and hence declutch a bit the worlding power of modeling that keeps both math and so-called plants in place. It is how "so-called" operates as a disclaimer, and thereby opens up possibility for the Ivies to make a difference. *Item 120: Simulated dendrochronology for demographics?* points at the need for eccentric desired life-patterns. Once we accept the limits of representation, visualizations of de-centralization, un-balancing and crookedness might make space for complexity.

Nobody really believes that managing plantations through AI is beyond violence, that so-called plants can be generated, that fugitives should be separated from leaves in the wind. In our technocultures of critique, it is not rare at all to share the views on "of course, those techniques are not neutral". Nevertheless, after studying the tricks and tips of volumetrics (from biomedicine, to mining, to sports or to court), we understood that once these complex worlds entangle with computation, the normalized assumptions of Cartesian optimization start to dominate and overrule. The cases we keep in the Possible Bodies inventory are each rather banal, far from exceptional and even everyday. They show that volumetrics are embedded in very mundane situations, but once folded into computation, concerns are easily dismissed. It shows the monocultural power of the probable, as a seemingly non-violent regulator of that what is predictable and therefore proportional, reasonable and efficient. The probable is an adjective turned into a noun, a world oriented by probabilistic vectors, in the socioeconomic sense of the "normal". We are committed to heightening sensibility for the actual violence of such normality, in order to start considering variable forms of opening up cracks for computational cultures that flourish by and for other means. By keeping complexity close, the possible becomes doable.

Notes

1. ↑ Fang Fang Gao et al., "Multi-Class Fruit-on-Plant Detection for Apple in SNAP System Using Faster R-CNN," *Computers and Electronics in Agriculture* 176 (2020).

2. ↑ Eckart Lange, Sigrid Hehl-Lange, "Integrating 3D Visualisation in Landscape Design and Environmental Planning," *GAIA Ecological Perspectives for Science and Society* 15, no. 3 (2006).

3. ↑ Dieter Fritsch and Martin Kada, "Visualisation Using Game Engines," *Archiwum ISPRS* 35. B5 (2004).

4. ↑ D.S. Pavithra and M.S. Srinath, "GSM based automatic irrigation control system for efficient use of resources and crop planning by using an Android mobile," *IOSR Journal of Mechanical and Civil Engineering* 11, no. 4 (August 2014): 49–55, https://doi.org/DOI:10.9790/16 84-11414955.

5. ↑ "RooT rak" (University of Nottingham, Computer Vision Laboratory), accessed April 1, 2021, https://www.nottingham.ac.uk/research/groups/cvl/software/rootrak.aspx.

6. ↑ Possible Bodies, "The Possible Bodies Inventory," 2022, https://possiblebodies.constantvzw.org.

7. ↑ Isabelle Stengers, "Introductory Notes on an Ecology of Practices," *Cultural Studies Review* 11, no. 1 (2005): 183–96.

8. ↑ Data-base-based cosmologies are modes of understanding and representation which are conformed by the material and imaginary constraints of fields, rows and columns.

9. ↑ See also: Jara Rocha and Femke Snelting, "Disorientation and Its Aftermath," in this book.

10. ↑ "Informatics of domination" is a term coined by Donna Haraway to refer to an emerging techno-social world order due to the transformation of power forms. Donna Haraway, "A Manifesto for Cyborgs: Science, technology, and socialist feminism in the 1980s," *Socialist Review*, no. 80 (1985): 985.

11. ↑ "Post-exotic" is a term coined by Livia Alga: "Nostalgia is exotic, memory is postexotic." Livia Alga, "there is always someone coming from the south," *Postesotica*, weblog, accessed October 6, 2020, http://postesotica.blogspot.com/p/blog-page_33.html/.

12. ↑ The industrial continuum of 3D has been a key figuration for the Possible Bodies research process, see Possible Bodies, "Item 074: The Continuum", *The Possible Bodies Inventory*, 2017.

13. ↑ María Puig Bellacasa, *Matters of Care: Speculative Ethics in More than Human Worlds* (Minneapolis: University of Minnesota Press, 2017).

14. ↑ "There are no solutions; there is only the ongoing practice of being open and alive to each meeting, each intra-action, so that we might use our ability to respond, our responsibility, to help awaken, to breathe life into ever new possibilities for living justly." Karen Barad, "Preface and Acknowledgements," in *Meeting the Universe Halfway: Quantum Physics and the Entanglement of Matter and Meaning* (Durham and London: Duke University Press, 2007), x.

15. ↑ Stefan Mairhofer, "Segmentation of a Tomato Root from Clay Loam Using RooTrak" (RooTrak 0.3.2, September 21, 2015), https://sourceforge.net/projects/rootrak/https://sourceforge.net/projects/rootrak/.

16. ↑ Rafael Sánchez-Mateos Paniagua, "En nuestros jardines se preparan bosques" (MUSAC, 2012). Translated by the authors.

17. ↑ Stefan Mairhofer et al., "RooTrak: Automated Recovery of Three-Dimensional Plant Root Architecture in Soil from X-Ray Microcomputed Tomography Images Using Visual Tracking," *Plant Physiology* 158, no. 2 (February 2012).

18. ↑ Pascale Barret, *This Obscure Side of Sweetness Is Waiting to Blossom* (TEMI Edition, South Korea, 2017), http://www.pascalebarret.com/projects/this-obscure-side-of-sweetness-is-waiting-to-blossom/.

19. ↑ See for another discussion of *Item 033*: Helen V. Pritchard, "Clumsy Volumetrics," in this book.

20. ↑ "Representation raised to the nth power does not disrupt the geometry that holds object and subject at a distance as the very condition for knowledge's possibility." Barad, "Diffractions: Differences, Contingencies, and Entanglements That Matter," *Meeting the Universe Halfway*, 88.

21. ↑ This perspective has been practiced with diverse sensibilities by authors as different as Zach Blas (*paranodal spaces*), Karen Barad ("What is the Measure of Nothingness?") or Gloria Anzaldúa (*Nepantlas*).

22. ↑ Mel Y. Chen, "Following Mercurial Affect," *Animacies: Biopolitics, Racial Mattering, and Queer Affect* (Durham and London: Duke University Press, 2012), 203.

23. ↑ "Circlusion is the antonym of penetration. What implications would it have, in naming, representing and performing switched-agency-practices, to push onto instead of pushing into?" Kym Ward feat. Possible Bodies, "Circluding," in this book.

24. ↑ Natasha Myers, "From the Anthropocene to the Planthroposcene: Designing Gardens for Plant/People Involution," *History and Anthropology* 18 n. 3 (2017).

25. ↑ Anna L. Tsing, "The Problem of Scale," *The Mushroom at the End of the World: On the Possibility of Life in Capitalist Ruins* (Princeton: Princeton University Press, 2015).

26. ↑ Adam Papp et al., "Automatic Annotation of Hyperspectral Images and Spectral Signal Classification of People and Vehicles in Areas of Dense Vegetation with Deep Learning," *Remote Sensing* 12, no. 13 (2020): 2111, https://doi.org/10.3390/rs12132111.

27. ↑ "FOLDOUT: Through Foliage Detection in the Inner and Outermost Regions of the EU" (Presentation flyer for the Mediterranean Security Event 2019), accessed October 6, 2020, https://mse2019.kemea-research.gr/wp-content/uploads/2019/11/FOLDOUT_AKriechbaum.pdf.

28. ↑ See diagram included in the second periodic project report for FOLDOUT on the Cordis EU Research results webpage. "Periodic Reporting for Period 2 - FOLDOUT (Through-Foliage Detection, Including in the Outermost Regions of the EU)," accessed December 6, 2021, https://cordis.europa.eu/project/id/787021/reporting.

29. ↑ R. Kapoor et al., "UWB Radar Detection of Targets in Foliage Using Alpha-Stable Clutter Models," *IEEE Transactions on Aerospace and Electronic Systems* 35, no. 3 (1999): 819–34.

30. ↑ "Frontex Programme of Work 2013," accessed October 6, 2020, https://www.statewatch.org/media/documents/observatories_files/frontex_observatory/Frontex%20Work%20Programme%202013.pdf.

31. ↑ Possible Bodies, "Item 122: So-Called Plants. Performance at Nepantlas #3." (Curated by Daphne Dragona at Akademie Schloss Solitude), accessed October 6, 2020, https://www.akademie-solitude.de/en/event/nepantlas-03/.

32. ↑ Image reproduced from: Image José Paulo Molin, Lucas Rios Amaral, and André Freitas Colaço, *Agricultura de precisão* (Sao Paulo: Oficina de textos, 2015).

33. ↑ Abelardo Gil-Fournier, "Earth Constellations: Agrarian Units and the Topological Partition of Space," *Geospatial Memory*, Media Theory 2, no. 1 (2018).

34. ↑ Gil-Fournier.

35. ↑ Abelardo Gil-Fournier, "An Earthology of Moving Landforms," 2017, http://abelardogfournier.org/works/earthology.html.

36. ↑ Gil-Fournier, "Earth Constellations: Agrarian Units and the Topological Partition of Space."

37. ↑ "Sawyer Seminar: Interrogating the Plantationocene," 2019, https://humanities.wisc.edu/research/plantationocene.

38. ↑ Jara Rocha and Femke Snelting, "The Industrial Continuum of 3D," in this book.

39. ↑ "A tool allowing a virtual ivy to grow in your 3d world," Thomas Luft, "An Ivy Generator," accessed November 1, 2021, http://graphics.uni-konstanz.de/~luft/ivy_generator/.

40. ↑ "Blender 2.92 Reference Manual," accessed April 11, 2021, https://docs.blender.org/manual/en/latest/.

41. ↑ Luft, "An Ivy Generator," 2007.

42. ↑ *Symbiogenesis* is a term that Lynn Margulis uses to refer to the crucial role of symbiosis in major evolutionary innovations. It literally means "becoming by living together." Lynn Margulis, "Genetic and evolutionary consequences of symbiosis," *Experimental Parasitology* 39, no. 2 (April 1976): 277–349.

43. ↑ Maddie Burakoff, "Decoding the Mathematical Secrets of Plants' Stunning Leaf Patterns," *Smithsonian Magazine*, June 6, 2019.

44. ↑ Jara Rocha and Femke Snelting, "Invasive Imaginatons and Its Agential Cuts," in this book.

45. ↑ Luft, "An Ivy Generator," 2007.

46. ↑ M. Grabner, D. Salaberger, and T. Okochi, "The Need of High Resolution μ-X-Ray CT in Dendrochronology and in Wood Identification," *Proceedings of 6th International Symposium on Image and Signal Processing and Analysis*, 2009.

47. ↑ "Laboratory Procedures: Basic Dendrochronology Procedures of the Cornell Tree-Ring Laboratory" (Cornell Tree-Ring Laboratory), accessed April 1, 2021, https://dendro.cornell.edu/whatisdendro.php.

48. ↑ Pedro Cruz et al., "Simulated Dendrochronology of U.S. Immigration," accessed April 1, 2020, https://web.northeastern.edu/naturalizing-immigration-dataviz/.

49. ↑ "Information Is Beautiful Awards," 2018, https://www.informationisbeautifulawards.com/showcase?action=index&award=2018&controller=showcase&page=1&pcategory=most-beautiful&type=awards.

50. ↑ Rebecca Cuningham, "Is Information Beautiful?," July 22, 2019, https://fakeflamenco.com/2019/07/22/is-information-beautiful/.

51. ↑ Denise Ferreira da Silva and Arjuna Neumann, "4 Waters: Deep Implicancy" (Images festival, 2019), http://archive.gallerytpw.ca/wp-content/uploads/2019/03/Arjuna-Denise-web-ready.pdf.

52. ↑ Bini Adamczak, "On Circlusion," *Mask Magazine*, 2016, http://www.maskmagazine.com/the-mommy-issue/sex/circlusion.

53. ↑ Karen Barad, "Nature's Queer Performativity*," ed. Køn Kvinder and Forskning, 2012.

Depths and Densities: Accidented and dissonant spacetimes

Open Boundary Conditions: A grid for intensive study

Kym Ward

	Watery Columns	Spongy Model Edges	Squints & True Colours
	CTD	FVCOM	MATLAB
Expanded old-school **FSTS** **Patronage / gender** **Social constructivist** **Who & where**	Challenger & colonialism	Accuracy & patronage 'good enough' measurements	'Color carries the responsibility of honesty' moral relativism
Measurements that matter **new materialisms** **agential cut**	Sammler Datum	Isometric net Cuts that divide problematics in data science — atmosphere model and scales of comparison	semiotics of color rainbow deception
Gestationality - speculative **life/non/life (problematizing distinction)** **phenomenological Relates to Scientific Prediction**	Wax non-life collection Neimanis	Biological model and life integration Cosmos as a technological system	Intuition for meaning of color map is natureculture data-vis as warnings not celebrations, exhaustion

"I think that perhaps there is importance in starting various forms of intensive learning and intensive study", Kym Ward explains when we ask about the grid that she devised to research Open Boundary Conditions. Kym works at Bidston Observatory Artistic Research Centre in Liverpool, a place that has historically been occupied by different research-led organizations — up to now, predominantly in the Natural Earth Sciences.[1] Originally built to measure time, latitude and the declination of the stars, in later iterations employees worked with meteorological, tidal and other marine data. Following this lineage from astronomical observation, to maritime scoping and charting, she became interested in the techno-political history of tidal prediction and started to study together with researchers from the National Oceanography Centre (NOC). In the following transcript, Kym explains us what is at stake in this work, and how it is structured.

An area of interest that needs focus

In the models that are used to run massive data sets, to do predictions for earth sciences or for meteorology or oceanography, there is an area of interest that needs to be focused on, because you can't collect and process all data. For example, if you're trying to figure out what waves will occur in a seascape, you need some edges to the object that you're looking at.

The issue with creating edges is that they just stop, that they make something finite, and things are often not finite. Waves have no edges and they don't just end. So, if you're trying to figure out different conditions for one area, a year in advance, you are going to have to figure out what comes in and what goes out of this imaginary realm. This is why you need what are called "open boundary conditions": the mathematics that are applied to hundreds of sets of variables that create the *outside* of that model in order for it to run.

There are a lot of different ways to create outside boundary conditions, and there are various kinds of equations that, in all honesty, are above my head. There are differential equations depending on what your object is, and if you're looking at waves, then you will use elliptic and hyperbolic equations.

The issue comes when you need to run two different kinds of data sets. You need to understand what wind is going to do to waves, for example. And if you need to know that, you are going to involve both the ocean model and the atmosphere model, which are on some level incompatible. The atmosphere model has many more data points than the ocean, something like at a ratio of 1000 to 1. What that means is that it is so much more fine grained than the ocean model, so they cannot simply be run together, for every time that there is one step of the ocean model, there is a thousand steps for the atmosphere model to run through. The open boundary conditions need to provide the sets of conditions that will allow for these models to be integrated at massively different scales. That is one example.

This term, "open boundary conditions", makes sense to me, because of the gathering and gleaning that I have been doing across different disciplines, knowing that the vocabularies and discipline-specific words I am using will be warped, and perhaps not have the same equations applied to them. But coming from critical media theory, or philosophy of technology, and then moving to applied sciences is going to produce some interesting differences in timescales and steps. The reason I'm talking about this at all, is that I landed at Bidston Observatory Artistic Research Centre, and this was formerly a place for astronomical observation. From astronomical observation it moved to tidal research and then prediction and charting. The history of the observatory as a part of the artistic research centre, which it is now, leads you to the kinds of data visualizations that are produced by modeling and data collection, and the discipline of oceanography as a whole.

Modelling Waves and Swerves

Modelling Waves and Swerves started off as a dusty scrabble around the basements.[2] I was excited to find original IBM 1130 data punch cards, which had been used in tidal prediction. But this soon turned into scratching my head over the harmonic calculations of tidal prediction machines, and I needed more help to understand these. And so, with collaborators, we set up *Modelling Waves and Swerves* — an ongoing series of weekend work sessions. In our initial call-out, we beckoned to "marine data modellers, tired oceanographers, software critics and people concerned with the politics of predictive visualizations". The tiredness was not a typo — it was intended as a mode of approach, of care, for the limits of a discipline; and to navigate between the steps of data collection, prediction and dispersal of climate change data. Repetitive conclusions of ocean warming and sea level rising are regularly released, and when these meet the reception from wider publics, which can sometimes at best be described as indifferent, surely scientists must be a little weary?

So these work sessions take place in the observatory, which was formerly occupied by the National Oceanography Centre (NOC), and sits just outside of Liverpool, in the UK. The group looks at current and historical processes of data collection, assimilation and computational modelling of oceanographic data sets, on the way to their visual outputs — and these chronologically range from ink blotted wavy lines on a drum of paper, to hyper-real 3D renderings.

The types of data visualizations we find now, are 3D ocean current models, or colour variated global warming indices. If we are asking about the looseness of attachment between data visualization and energetic response, and why there is so little real response to those snippish heat stripes, then in an appeal to ethics and behavioural change, it might be useful to reexamine some methodologies of data science for their onto-epistemological grounds. This is the focus of "open boundary conditions".

One of the initial questions that the oceanographers asked us in these workshops, was why the visualizations they have been doing aren't being received in a way which creates real change, why there is a deadening of effects when they produce their outputs even though they come in beautiful color stripes. They come in swirling movements across the globe, something that quite clearly shows the warming, why you can see sea level rise on their cross-section maps. These are obviously worrying, and if we take them seriously, they pose existential threat.

I think there are a lot of artists and designers who would happily produce "better" visualizations, but you have to wonder what are the parameters of "better" in this case? More affective? Seemingly more "real"? In fact, what we're interested in is the steps to get to the visualizations in the first place. So, the collections of data, the running of models, and then the output.

A grid but not a monument

The first thing to note is the impossibility of conducting this kind of research alone: if it were important, it would be important to more people than me. So I'm not very precious about the grid that I have proposed. It's not a monument. I think that perhaps there is importance in starting various forms of intensive learning, intensive study, which I see there is also a desire for.

I haven't seen the desire for exploring and explaining the techno-logical back-end but I do see the desire for trying to get to grips with understanding oceanality and the ocean in an ecological sense. So I can see that there would be amazing possibilities for working with other people, in which you would hope that it wouldn't all be strug-gling with text. That it could find some visual form, that it could find some practical form, that it could find some performance form, work-ing in combination with the histories of science as they are, but also recombining to make other forms of knowledge. I would never have done this without the oceanographers and the data scientists. There is no possibility that I could have understood harmonic constants with-out a little bit of input.

Yes, it comes form a concern that by working with a critique of technological processes of oceanography, towards data visualisation, I'm only deconstructing the different inheritances of Modernity. For example, in looking at biopower through affect theory, looking at the way that color affects the regulation of the body and its response. Or looking at it through a criticism and awareness of colonial history, and how that's built the technologies in both extractivist and utilitar-ian ways. There's a legitimacy in doing that, but it doesn't create any kind of constructive conversation with anyone that I've been working with- with oceanographers, with data scientists. It does create pro-ductive conversations with philosophers but that might not reach any conclusion.

My suspicion was that certain discourses that are happening in feminist science studies, in new materialisms and in feminist phe-nomenology could add to an understanding that in the end, a color stripe might not make that much difference, or create inaction. To do that, rather than to just open some books and read some pages, I thought that it would be more invested and involved, and careful and considerate and honest, and also confused, to take some objects and try to talk these through discourses and questions via those objects. So, I picked three.

Watery Columns: The CTD monitor

The first example I picked was a CTD Monitor. CTD Monitor is a metal instrument which gets dropped down from an amazing buoy. There will be 10 or 12 CTDs which are arranged in a ring, and they get dropped, and sink to the bottom of the ocean. And then at some point, on a timer, they are released, and they will rise. And as they rise, their little metal mouths will open up and grab a gulp of sea water at a particular level. The mouths will close and they will proceed to the top and at some point they will be collected and this happens over a certain time period. Its testing for salinity, its testing for temperature, its testing for depth. Salinity is measured by conductivity and hydrostatic pressure I think.

This logic follows long history of the way that the seascape is carved up, which the CTD instruments will rise through. Originally, it would have been a hemp rope, weighted with lead, which would be dropped from the side of a ship, As it drops, it runs through the hands of the sailors. There are knots on the rope, and each knot represents a fathom, and the fathoms are called out, and someone marks them with a quill pen.

Through the architecture of Modernity, oceanography has the way of imagining the sea as a column. The sea is a very unstriated space that is imagined as an unchanging space. Even until today, this is how information is collected. Even the more unusual forms of data collection, such as the mini CTDs that are glued onto the heads of seals (a lot of the arctic data is from different seals who swim around). There is a GPS attached to it, and it still logged even though the seal is still swimming happily with that thing glued to its head. The sea is still divided up into a grid, at a certain depth, what is the salinity, temperature and conductivity, for example.

So, even when sea mammals are put to work doing scientific investigation, and this investigation is then recalibrated into what is fundamentally a giant technological system formed on axes, really. It really brings home the quite strict ontological ground for sea exploration, and the types of relationality that happen in a vast expanse of many different types of sea lives, and many different kinds of waters. Under sea vents, tectonic plates, underwater volcanoes, ecologies which are then being programmed into fundamentally the same model. The data are being used not to explore something different, but to expand Western knowledges along an axis.

Spongy Model Edges: FVCOM

Another way that the seascape is absurdly chopped or divided from its messiness and never-ending movement is the construction of maritime boundaries, which are basically virtual objects in the sea, which are carved up by what is a nation state, by what is landmass. They are geopolitical artifacts. For example, since the late 1700s, at one of the points in the Americas, at Saint Martha's Bay, the sea is recorded all the way down that coast, over the period of a year, and the mean sea-level is found. It's a mean sea-level, because tides go up and down, there are semi-diurnal tides, there are diurnal tides, there are mixed tides. There's waves! There are still sea movements that are foxing oceanographers. But in any event, the sea was averaged, there was highest point, the lowest point and the mean sea level was used to construct a zero, a datum. And from this point you start to measure mountains, upwards. How many kilometers above the sea is, how can you measure the sea? You measure it from the average of the sea. It's absurd, but it's also the globally agreed protocol.

So what happens when you introduce climate change into this phenomenon is that mountains start shrinking because sea levels are rising. It has sociological, geological, urban planning, planning appli-cations, which are in end effects political. What is classified as a dis-appearing island, or a non-disappearing island becomes ratified.

FVCom is one of many multiple models that are used as a coordi-nate system. The example I gave earlier is just one example of data that is collected: salinity, temperature, depth, and obviously there are billions of data points that are also collected along rivers, along the coastline, and within the sea. One of the interesting things about how data is collected is that the nodes of data collection are very tightly packed around the coastlines, near rivers, and they are done on an isomorphic net, so it's a triangular grid system that can be scaled. It can be expanded or contracted depending how close you want to zoom into that particular part of ocean, or coastline. And as you move out to sea, the grid gets a lot bigger. So the point at which the data is collected is averaged so that the data can run. And way out it into the middle of the ocean, you might have a two kilometer or three mile point between each of those corners of the triangle of this net which, anywhere between this node, gets averaged. Whereas at the coastline, you'll have much tighter data, and the net will be in centimeters, or meters, not in miles.

So FVCom is one of the many models, called "the ocean model" that we've been looking into. All of these models begin in the late '60s, early '70s and onward, they've been developed along the way in the intervening years and they take on more data points. What was ini-tially not understood as being part of the ocean will then form one of the later models, for example, the biological model which is made of tiny life forms, phytoplankton and zooplankton — that came later. I already talked a little bit about how the models overlap and sync with each other.

Sponginess is a term used to describe the boundary conditions where one massive model meets another massive model. The data which was collected to put into the model, if I describe it historically, one of the ways in which the process of modeling happens, is — someone takes measurements over the course of the coastline over a year, and the data is sent in. And the sheets of data that are sent in would be really grubby — they would perhaps be water sodden; but they were basic tabulations about the tide heights, the moon, the distances between waves. Different data like that. Before the advent of computers as we know them now, this information would be sent, in this way to Bidston Observatory, so that's my access point into this history. And then that data would be fundamentally programmed so that the height of the tides or the wavelength, or the effect of the moon, would be run through different differential equations, and then it would be assigned a value. The value would be put into a tidal prediction machine. This machine was made of metal, with 42 brass discs. A band ran in-between these discs, each of the discs had a different name — for example, $m2$ was the moon. And these discs would move up and down on arms. What was produced at the end of this computation- placed onto a roll of paper that was also onto a spinning drum by an arm, attached on one end with an ink pot, and the pen at the other which would draw out the harmonics — a wave. This wave was a prediction for next years tides.

The tidal prediction machines around the time of the Second World War could do one year's worth of predictions in one day. Different places around the world would send in their tidal calculations and they would receive back the predictions for the year, saying at what time what tide what height. The different harmonic constants, as they were called, that were run through the tidal prediction machines, they find themselves still in the predictions nowadays. They've been massively updated, and there are obviously so many more data points- but you can still find them in how FVCom works.

One of the interesting things that happen in-between data collection, human error, different calculations and output, is that sometimes you get an output that does not resemble a harmonic — it doesn't resemble a wave form. It needs to be smoothed. At that time, in order to correct it, it was simply rubbed out and drawn on with a pencil. The computers in the 1930s (the women who operated the machines were called computers), had partners — the "smoother", whose task it was to correct the prediction blip. I see that there is a connection between the isomorphic grid with the averages in the middle of the sea, and the job of the "smoother". They are both attempts to speak to what is legitimate accuracy.

One of the strands of research that I've been doing was helped a lot by a feminist science and technology scholar, Anna Carlsson-Hyslop, and she wrote a paper on Doodson, one of the previous directors of the observatory.[3] He was doing a lot of work on tidal prediction. She traces a line from his conscientious objection in the First World War to his subsequent work on aircraft ballistics. So while he doesn't want to go to war, he doesn't want to fight, he won't go, he is conscripted to do mathematical scientific research because he is good at math, to do calculations on the trajectory of bombs, instead of going to war. As a part of this work that he did, he developed a way of looking at the arc of a missile using differential equations.

Carlsson-Hyslop writes about the interaction between patronage and what is an accurate calculation. In order for these calculations to be done, somebody's got to pay for them. Doodson is receiving a wage, but he also knows that there are "good enough" calculations for this set of conditions. When we think of the lineage of modeling, the impetus is to become more and more accurate. But its super helpful to keep in mind that there is a difference between *accuracy* and *legitimacy*. The necessity for accuracy supposedly makes it more legitimate, however, it doesn't correlate from a feminist science point of view.

I'm just trying to figure out why I thought that the depths were denser. Obviously they are because there is more life there. The amassed points of interest are not the same as organic life. The surface of the water is more recordable, visible, datafiable. The depths are unknown. I think I was trying to make a link to what superficial means... like does it mean whether there's something productive in a literary sense. Superficial is able to be captured a lot easier.

Squints & True Colours: CM Ocean

The third object of study is called CM Ocean. It's a programming software that is running MATLAB, in order to output the data which has then been run in the model. It is a visualization that would run alongside, and produce varying different scales of data via color. So there's a lot of different programs which can turn ocean data into color, like heat stripes, water warming, sea warming, water level rise, salinity... lots of different kinds of data.

We started off this journey speaking about why visualization don't produce effect when they have to do with existential questions like Climate Change. So it makes sense to talk about CM Ocean.

The data that is transformed into these visualizations are numerical, it's quantity. And then they are translated into a scale that is absolutely not numerical, and are very subjective in terms of its reception. The aim of CM Ocean is to desubjectify and to make colour scientific. It is quite a task, which is surprising that a group would take it on. But CM Ocean is funded by BP, a multinational oil and gas company, and funded by George Bush. Its not that necessarily this has a one-on-one effect. But it's obvious, and worth noting that an oil company and the Texas Government would like to have a regulated way of understanding the contents of the ocean.

The second thing is that the subjectivity of color is aimed for regulation, which bypasses things like taste. It bypasses any kind of physiological reception. I was thinking that perhaps the expectation that color can be reproducible, that it can be accurate, that it can correctly represent numerical data, that it can't be divorced from numericizing color in the first place, the attributions of CMYK and RGB. If color is printed, it is different to if it's on a screen. There are so many unworkables to this method, if you think about it. But the belief is that its good color usage carries the responsibility of honesty. So, to use colors in an honest way is the responsibility of the scientist. But what is honesty in color representation of data points? Its previous iteration, called JETS, is supposedly not so accurate, not so precise because it has the movements through the color scale with arbitrary weights. So, this has you thinking that there's a density of whatever it is you're looking for in the ocean because this particular part of the color scale is more dense to you, to the reception of the eye. Dark purple rather than light yellow might misrepresent the density of the object in question, but you would never know that, because this perceived symbolism is skewed. The gradient of the color has to accelerate and decelerate but it might not do that at the scale of the numerical values have on the back-end. It might be that it looks like it's getting warmer quickly, but it depending on how this color scale is being applied, it could completely skew the numerical results that you've run your model for.

It's also worth saying, that these models are hugely energy expensive, and take around forty days to run. The step from programming to output, massive amounts of electricity are used and the possibility for it to go wrong are quite large. If so, you would have to start again and try to recalculate. As I mentioned at the start of this conversation, if we look at these instances in the process of data collection to output, solely in a critical mode, then we fail in a remarkable way: the ocean, its inhabitants, what is life and what sustains us on this planet, is still and always our object of study. We need to propose other methods of working together, of offering feedback, which differently separate our object, or work with separability itself. The grid-not-monument we're working with here, is a try towards this.

Datum point installed in the basement of Bidston Observatory, Kym Ward, 2021

Frame: Expanded old school

I want to try to think through these three cases in an expanded, old-school, social-constructivist feminist way where you would think about where that object is being produced, who produced it, how does it have an effect on and are there any, what are the linguistic and semiotic exchanges that take place because this technology has been built in this, and has been used by these people on these people. On these bodies, by bodies I mean the ocean, the body of water.

It is about naming where and when something has been produced, in order to properly understand the limitations of its production, about making clear the ramifications of who and not resorting to default "I" or displaced I of objectivity.

Frame: Measurements that matter

The second frame is to use some of the work that has been done over the last 10 to 20 years on new materialism, to try to think about how for the fact that all of these objects measure in different ways, they produce matter in the way that they measure. So the CTD Monitor measures only X, it makes an apparatus which combines and makes the world in a certain way. Which is then, only just a tiny little data point which then is put into FV Com. It's difficult to talk about FV Com through new materialism, because it is such an object, but it can be done in a kind of reflective mode.

We tried quite hard in *Modeling Waves and Swerves*, to work this frame. It is possible, but it's much easier to look at one instrument than it is to look at a combination of instruments that form a massive instrument.

And also in the impossibility of retreat from a massive models that separate ocean life and atmosphere, for example. You need one of those models in order to have input on the data, but because they have already been divided in a certain way, you have to run with the implications of that. It is a lot easier when you go all the way out, but not when you are looking at FV Com and your looking at the back-end in order to understand as an oceanographer or a data scientist, thinking, "OK, what would the agential cut be?".

Frame: Gestationality

And the third strand, I call it "the feminist phenomenological", but it really comes from reading the work the of Astrida Neimanis, who wrote *Bodies of Water*.[4] In the book, she speaks to ontologic and ontologics, on the ontological of amniotics, and she is calling ontologic- not ontology which would deal with what "is" — but rather a who what when where how of commons of whatever it is we call more then human interlocutors. So, she speaks about amniotic in permeable open boundary membrane kind of ways. She is not only speaking about life that forms in the way of what she calls amniotes, life which forms in an amniotic sack, but she's also using it as a metaphor, as a fictional philosophical tool which is useful.

The reason that I had centered on this is why would feminist phenomenology have something to do with different modes of technical production of the ocean? She speaks to the water, different bodies of water that were along an evolutionary process, but also she speaks to them as a mode of reception and understanding and oneness with what is happening in the ocean. So it's a mode of understanding climate change, of potentially understanding sea warming. It has a lived bodily reality that we can connect to.

The second reason that I thought it would be worthwhile to walk down this path a little bit was because if your thinking about the ontologics of amniotics, you're also thinking about gestationality, and gestationality also makes sense when you're talking about predictions, ocean predictions. Because what, in the end, what this movement between data collection and running the models and producing the visualizations defines what is seen to be the ocean, and what is not seen to be the ocean, the contents of the ocean, the conditions of the ocean, the life of the ocean, what is not life in the ocean. And the kind of predictions that are accredited and valued by science are highly technologized predictions.

The idea of what gestationality does is that it posits that life could come, the possibility for life is there, but we don't know what kind of life will come and what it will look like. We don't have a clue of it, its on the move and its emergent but there is no form to it yet. And this is something that I find, compared to prediction and its vast technologies that I tried to describe, I find gestationality useful and very exciting.

Notes

1. ↑ "Bidston Observatory Artistic Research Centre (BOARC)," accessed October 20, 2021, http://www.bidston observatory.org.
2. ↑ Open call for "Modeling Swerves and Waves," accessed October 20, 2021, http://www.bidstonobservator y.org/?modelling_waves_swerves.
3. ↑ Anna Carlsson-Hyslop, *An Anatomy of storm surge science at Liverpool Tidal Institute 1919-1959:* *Forecasting, practices of calculation and patronage,* thesis submitted to the University of Manchester for the degree of Doctor of Philosophy in the Faculty of Life Sciences, 2010.
4. ↑ Astrida Neimanis, *Bodies of Water: Posthuman Feminist Phenomenology* (Edingburgh: Edingburgh University Press, 2017).

Depths and Densities:
A bugged report

Jara Rocha

Under the guise of a one-afternoon workshop at transmediale 2019, Possible Bodies proposed to collectively study open-source tools for geo-modelling while attending to the different regimes – of truth, of representation, of language or of political ideology – they operate within. It attempted to read those tools and a selection of texts in relation, with the plan of injecting some resistant vocabularies, misuses and/or f(r)ictions that could affect the extractivist bias embedded in the computation of earth's depths and densities.

The workshop *Depths and Densities* was populated by a mix of known companions and just-met participants (in total, a convergence of circa thirty voices), each bringing her own particular intensities regarding the tools, the theories, the vocabularies, and the urgencies placed upon the table. The discussions were recorded on the spot and transcribed later. This report cuts through a thick mass of written notes, transcriptions, and excerpted theoretical texts, sedimented along five vectorial provocations: *on the standardisation of time*, *on software vocabularies*, *on the activation of geontologies*, *on the computation of velocities*, and *on the techniques of 3D visualizations*. Each vectorial provocation was taken up by a sub-group of participants, who assumed the task of opening up a piece of Gplates, a free software tool and web portal for tectonic plate modeling. By holding close a technical feature, a forum, a tutorial, an interface etc. for a few hours, and tensioning these with some text matter from a reader pre-cooked by Helen V. Pritchard, Femke Snelting, and myself, Gplates worked as a catalyst for our conversations. Its community of developers would eventually become the deferred interlocutors of a report.[1]

The following cut was made to share a sample of that afternoon's eclectic dialogues in what could be transferred as a polyphonic bugged report. All text injections (in italics, on the right side) are quotes taken from the workshop's reader. All pieces following one already quoted belong to the same author, until the next quote in italics appears. All voices on the left emerged along the workshop's discussion, which was transcribed by Fanny Wendt Höjer.[2]

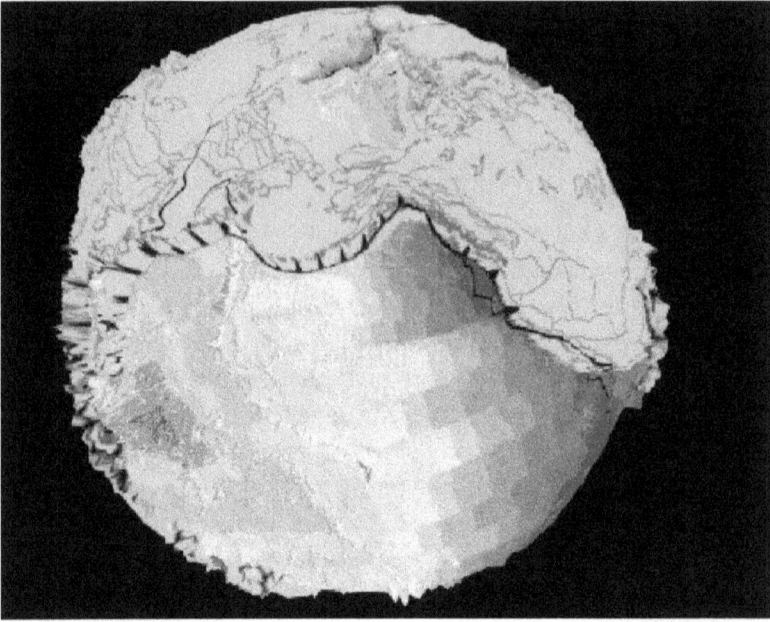

**First vectorial provocation,
on standardized time**

if multiple timescales are sedimented in contemporary software
environments used by geophysics, can fossil fuel extractivist practices
be understood as time-travelling practices?

*in these troubling times, there is an urgency to trouble time, to shake
it to its core, and to produce collective imaginaries that undo pervasive
conceptions of temporality.*[3]

this urgency is both new and not new

how is the end of time imagined, in a modelling sense?

we see discretely plotted colors

time isn't what it used to be

does the body of earth exist in the same timescale as you do?

or try and witness the whens otherwise

time tends to be limited to (and influenced by) the observer's perception but what are the material and semiotic conditions for another kind of time perception?

sedimented time and coexistence at ecologies of nothingness (aka voids)

> *voids are features that occur commonly in near-surface geophysical imaging. [...] However, voids are often misidentified. Some voids are missed, and other anomalous features are misinterpreted as voids, when in fact they are not. Compare them with real voids, and we determinate the differences based on incomplete data*[4]

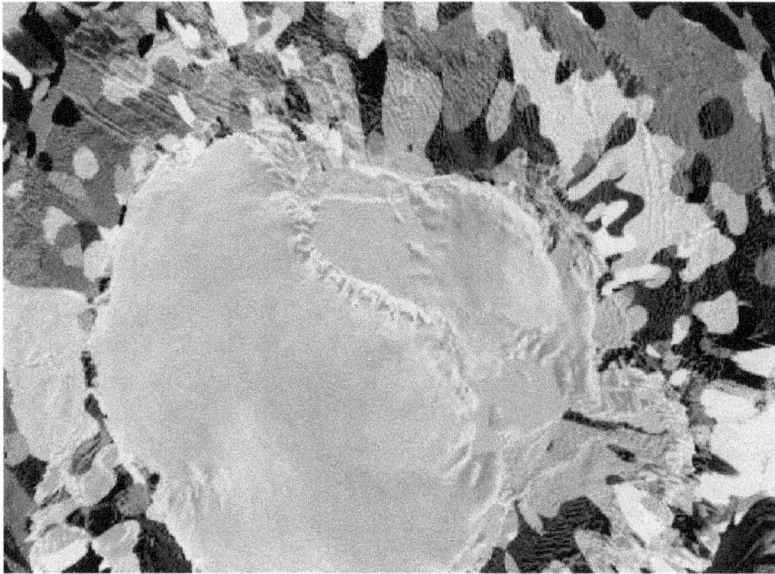

Second vectorial provocation, on software vocabularies

forging a differently fueled language of geology must provide a lexicon with which to attend the geotraumas

> *the endurance of a stony patience that doesn't forget love*[5]

user engagement with the earth through a 3D visualization software is based on metaphors like handling or grabbing *in the lexicon of geology that takes possession of people and places, delimiting the organization of existence, the refusal of such captivity makes a commons in the measure and pitch of the world, not the exclusive universality of the humanist subject*

you can still grab the earth: at Gplates a stable static earth is available for grabbing

a refusal to be delimited is found in the matter of the world and a home in its maroonage; "they wander as if they have no century, as if they can bound time... compasses whose directions tilt, skid off known maps"

also, the use of the verb "to grab" brings with it the history and practice of "land grabbing", land abuse and arbitrary actions of ownership and appropriation with correlated both dispossession by the taking of land, and environmental damage

but what if the earth grabs back?

there is a kind of reason that we will no longer accept tilting the axis of engagement within a geological optic and intimacy, the inhuman can be claimed as a different kind of resource than in its propertied colonial form—a gravitational form so extravagant, it defies gravity

if all the semantic network of Gplates is based on handling and grabbing as a key gestures in relation to the body of earth, a loss of agency and extractivist assumption slip in too smoothly, and too fast

forging a new language of geology must provide a lexicon with which to take apart the Anthropocene, a poetry to refashion a new epoch, a new geology that attends the the racialization of matter

most software platforms allow for no resistance, for no possible unavailability

the praxis of that aesthetic locates an insurgent geology

middle click and drag ¡la tierra para quien la trabaja![6]

*reconstituted in terms of agency for the present, for the end of this
world and the possibility of others, because the world is already
turning*

and what if the earth grabs back

the ghosts of geology rise

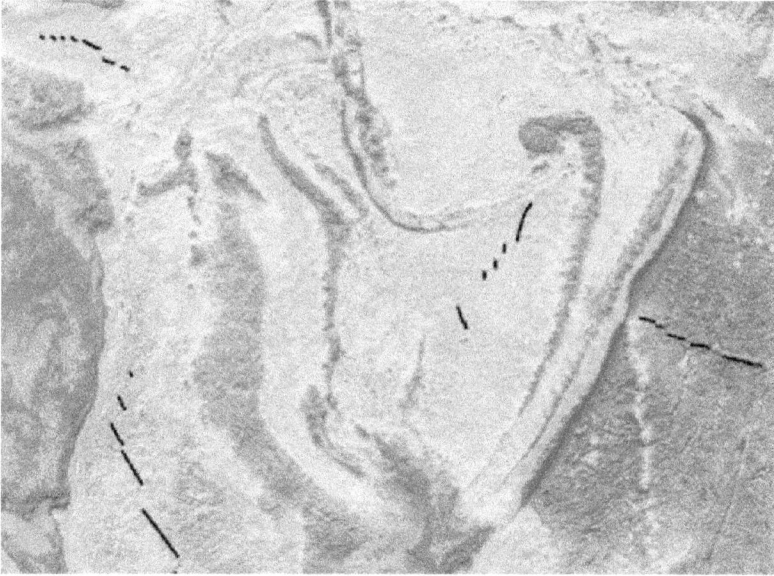

**Third vectorial provocation,
on the activation of geontologies**

we are all talking over each other like tectonic plates and strata

a time of the geos, of soullessness[7]

looking at what geology is implies a reconsideration of assumptions
of what life is

*the anthropos as just one element in the larger set of not merely
animal life but all Life as opposed to the state of original and radical
Nonlife*

minerals rocks plates

the vital in relation to the inert, the extinct in relation to the barren

cannot be separated from time

it is also clear that late liberal strategies for governing difference and markets also only work insofar as these distinctions are maintained

but where is the legend we could not read it

Life (Life{birth, growth, reproduction}v. Death) v. Nonlife

why this suspension subversion of the living

why this suspension subversion of the living

it is hardly an uncontroversial concept

otherwise the future will keep being missing but wait, the past is also missing the line goes back to 172 million years but earth is 4,5 billion years

the way data gets laid over particular shapes, how that comes to kind of operationalize particular makings and matterings of the world.[8]

a color-coded chronology is that tone the year of emergence or is it duration of collapse of merging

so kind of thinking through the technical and political questions of what is depth and what is density, how they shift depending on the situation they're operationalized within

a gradient of abstraction is being dangerously portrayed

the differences perhaps of the densities in geophysics to the densities in something like biomedical scanning, even though both might have tomographic processes

what is the skin of a body its density how is it colored?

density is not a fixed thing

but why?

we're interested in exploring these open questions; how these matter,
and how they matter in relation to things like surfaces and their
topologies, where there might be densities of power

a chroma chart would be appreciated

there's a kind of thickness in imaginaries of depth: the kind of
unknown or unreachable, the removed or the unremovable. But also
the kind of dark and morally crooked in bodies, in earth and in desires

like absolute dating of rocks you're alive, I'm alive/let's go

but other imaginations of depths in relation to both the earth or the
so-called body, or the body of the earth. In particular, the thinking with
the kind of writing from geo-philosophy and feminist technoscience,
which might suggest that we might tilt the axis of engagement

peel earth's skin the mantle

i think that's at heart of the Possible Bodies project as well, this tilting
of access to a different kind of optic

and peel it back where 4D is time and meets 5D uncertainty

to a different kind of intimacy

it does not peel back enough

think about the inhuman of earth surfaces, of tectonic plates, of
geological strata; they might have another possibility than the
proprietary colonial form, which often is the way it gets rendered
within things like the modelling tools for say the extraction of fossil
fuels or natural gas

Geontologies: the need of all bug reports

Fourth vectorial provocation,
on computing velocities

that is too linear, this is too straight

data has different densities and intensities and the effects and affects
of the single timeline make themselves visible

when specific intra-active technologies violently rendered real bodies,
they wondered about the see-through space-times that were left in the
dark[9]

leaving grey areas that show no data coverage

the crisis of presence that emerged with the computational turn was
shaped by the technocolonialism of turbocapitalism!

where is that information what is this superfiction

convoked from the dark inner space-times of the earth, the flesh and
the cosmos, particular [amodern] renderings evidence that real bodies
do not exist before being separated, cut and isolated.

whole parts of grey earth like you are making a cake you can put toppings on

grey means there is nothing such as a body of earth it is almost a void

they read, listened and gossiped with awkwardness, intensity and
urgency

earth used as a template for almost always fractured data

listen: there is a shaking surface, a cosmological inventory, hot breath
in the ear

zoom in this shaking surface and always find some cracks

the tool keeps wanting it to be presented as a whole the oneness of
earthness as in the oneness of humanness

there is a persistently imposing paradigm of wholeness and a preten-
sion of full resolution but a body becomes any body only if the whole
thing collapses

but when

[the soil] is no longer (or never was) the exclusive realm of technocrats
or geophysics experts

swipe it fast so much time in one swipe

it is almost rude

these are your new devices, dim and glossy

take your time scroll scroll scroll deeper

where poetic renderings start to (re)generate (just) social
imaginations

theres thens truths

let's collectively resonate against technologies

counting backwards and year zero does not stay

grab that time and

perhaps if you upgrade the software you can get extra time

that bring in trans∗feminist queer futures

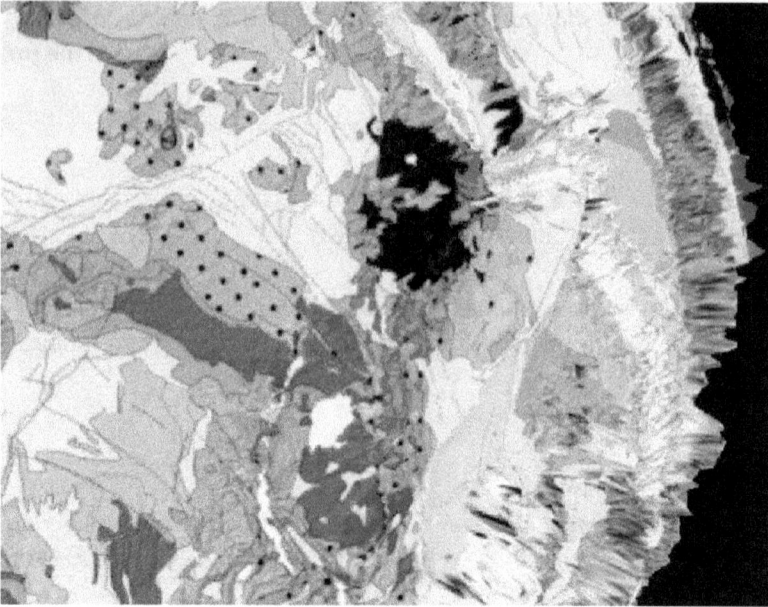

**Fifth vectorial provocation,
on the techniques of 3D volume visualization**

who is behind the proposers of the Mercator projection[10]

postcolonial or hegemonic structures of development[11]

who is behind one more eurocentric view of it

*"the centrality of mathematical and technological science... structured
by masculinist ideologies of domination and mastery"*

from 2D to 3D

such institutional, cultural, and scientific practices also affect glaciological knowledge

you are the camera!

Questions of who produces glaciological knowledge, and how such knowledge is used or shared, take on real implications when considered through feminist postcolonial science studies and feminist political ecology lenses

At Gplates you can replace the pole location grab the pole and drag it

indigenous accounts do not portray the ice as passive, to be measured and mastered

while time happens along a linear highlight of cascading data

folk glaciologies diversify the field of glaciology and subvert the hegemony of natural sciences

Gplates applies deep familiar metaphors like child plates

Of the Earth, the present subject of our scenarios, we can presuppose a single thing: it doesn't care about the questions we ask about it[12]

slide the zoom in and out of a data set of magnetic information

to speak of a world which is "prior" and "independent" without implying that it is "single" and "determinate": it encounters an earth which is very much "already composed" without it thereby being "already totalized"[13]

now

relocate

the pole

having "a stable identity" in relation to scientific study does not imply stasis or stability per se

slide

deeper down

smoothly

but how when where

but who what why

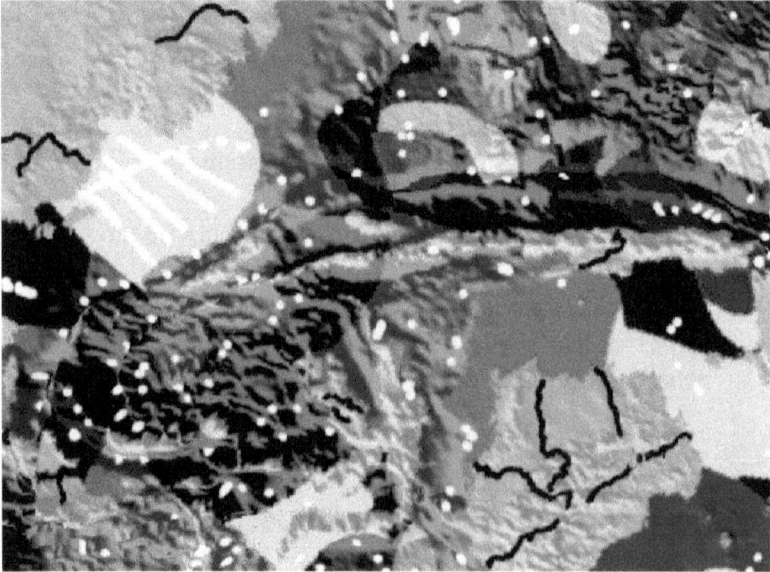

Notes

1. ↑ See for a continuation of these interlocutions, The Underground Division (Helen V. Pritchard, Jara Rocha, Femke Snelting), "We Have Always Been Geohackers," in this book.

2. ↑ See also: "Item 114: Earth Grabs Back," *The Possible Bodies Inventory*, 2019.

3. ↑ Karen Barad, "Troubling Time/s and Ecologies of Nothingness: on the im/possibilities of living and dying in the void," *New Formations 92: Posthuman Temporalities* (2018).

4. ↑ David C. Nobes, "Pitfalls to Avoid in Void Interpretation from Ground Penetrating Radar Imaging," *Interpretation* 6 (June 2018): 1-31. 10.1190/int-2018-0049.1.

5. ↑ Kathryn Yusoff, *A Billion Black Anthropocenes or None* (Minneapolis: University of Minnesota Press, 2018).

6. ↑ Emiliano Zapata (c.1911).

7. ↑ Elizabeth A. Povinelli, *Geontologies: A requiem to late liberalism* (Durham: Duke University Press, 2016).

8. ↑ Excerpts from Helen V. Pritchard's oral introduction to the workshop.

9. ↑ Possible Bodies feat. Helen Pritchard, "Ultrasonic Dreams of Aclinical Renderings," in this book.

10. ↑ "Mercator Projection," Wikipedia, https://en.wikipedia.org/wiki/Mercator_projection.

11. ↑ Mark Carey, M. Jackson, Alessandro Antonello and Jaclyn Rushing, "Glaciers, Gender, and Science: A feminist glaciology framework for global environmental change research," *Progress in Human Geography 40, no. 6 (2016)*: 770-793.

12. ↑ Isabelle Stengers, *The Invention of Modern Science* (Minneapolis: University of Minnesota Press, 2000).

13. ↑ Nigel Clark, "Inhuman Nature: Sociable Life on a Dynamic Planet," *Theory, Culture & Society* (2011): 38-39.

We Have Always Been Geohackers

The Underground Division
(Helen V. Pritchard, Jara Rocha, Femke Snelting)

The Anthropocene should go in a bug report, in the mother of all bug reports. It is hardly an uncontroversial concept.[1]

Time:	65.00 ⌄ Ma	▷	◁◁	◁◁	▷▷	═══════════╪══════

Detail of Gplates main interface (timeslider)

Triggered by a lack of trans∗feminist experiments with volumetric geocomputation techniques and the necessity to engage with a counterhistory of geologic relations, the Underground Division took a leap of both scale and time, which implicated a jump from inquiries into the field of body politics to considerations of geopolitics. Together with a group of companions participating in *Depths and Densities*, a workshop in the context of transmediale festival 2019, we moved from individual somatic corporealities (or zoologically-recognized organisms) towards the so-called *body of the earth*.[2] Our trans∗feminist vector was sharpened by queer and antiracist sensibilities, and oriented towards (but not limited to) trans∗generational, trans∗media, trans∗disciplinary, trans∗geopolitical, trans∗expertise, and trans∗genealogical concerns.[3] Collectively we explored the volumetric renderings of the so-called earth and how they are made operative by geocomputation, where geocomputation refers to the computational processes that measure, quantify, historicize, visualize, predict, classify, model, and tell stories of spatial and temporal geologic processes. We invited participants to collectively report bugs found through/on Gplates, a free software tool and web portal for tectonic plate modeling.[4] What emerged in the bug reporting was the urgent need to generate figures and operations that are not dependent on the expertise of technocrats, experts, or technoscience. As a way into this, in this chapter we mobilize the methodological figures of *disobedient bug reporting* and *disobedient action research* to ask — what affirmative forms of responsibility-taking might be possible through taking up these figures within the processes and practices of volumetric geocomputation? *The Depths and Densities* workshop triangulated Gplates' visions of the earth with critical software and interface analysis, poetics, and theoretical text materials. Working through Gplates is a consideration of volumetric regimes as world building practices. For us, it was in part a response to Yusoff's call for "a need to examine the epistemological framings and categorizations that produce the material and discursive world building through geology in both its historical and present forms".[5] In this way, we attended to the material-discursive amalgam of Gplates: the different regimes of truth, histories, representation, language, and political ideology that operate upon it.[6] While staying close to an approach that holds that the underground is no longer (or never was) the exclusive realm of technocrats or geophysics experts, this text is based on discussions and reflections that flowed from the workshop.

Volumetric regimes

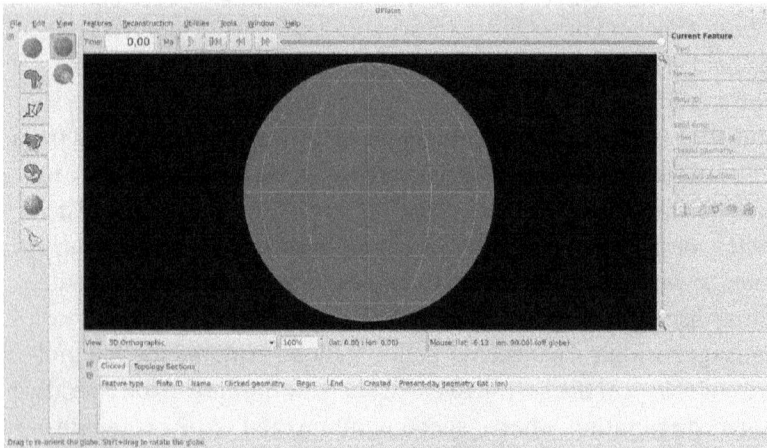

Gplates interface before loading geodata (grey earth)

Geomodelling software contributes to technocolonial subsurface exploration and extraction by enlisting, among other things, geophysics stratigraphy, diagenesis, paleoclimatology, structural geology, and sedimentology combined with computational techniques and paradigms for acquiring and rendering volumetric data. Following the *industrial continuum of 3D*, the same techniques and manners that power subsurface exploration are operationalized within other domains, such as, for example biomedical imaging, entertainment industries, and border policing.[7] In that sense, jumps in scale from individual somatic corporealities to the so-called body of the earth is daily business for the industries of volumetrics.

We chose to work with Gplates because it is a software platform that emerges from a complex web of academic, corporate, and software interests that allows communities of geophysicists to reconstruct, visualize, and manipulate complex plate-tectonic data-sets. For users with other types of expertise, Gplates provides a web portal with the possibility of on-the-fly rendering of selected data sets, such as LiDAR Data, Paleomagnetic Data, and Gravity Anomalies.[8] The software is published under a general public license which means its code is legally available for inspection, distribution, and collaboration.

According to its own description, Gplates offers a novel combination of interactive plate-tectonic reconstructions, geographic information system (GIS) functionality and raster data visualization. GPlates enables both the visualization and the manipulation of plate-tectonic reconstructions and associated data through geological time.[9]

The application is developed by a global consortium of academic research institutions situated in geological and planetary sciences. EarthByte, the consortium's leading partner, is an "international center of excellence and industry partners" whose large team is formed by students, researchers in oceanography and geology, and employees assigned to the project by companies, such as Shell, Chevron, and Statoil.[10] Gplates implements its own native file format, the Gplates Markup Language (GPML), in order to combine and visualize public data-sets from various resources, and to render them onto the basic shape of a gray globe.[11] A horizontal timeline invites users to animate tectonic plate movement seamlessly forwards and backwards over geological time.

As software was downloaded during the workshop, knowledge and relations commingled, and soon, fifteen laptops were displaying the Gplates portal. Together we imagined resistant vocabularies, creative misuses and/or plausible f(r)ictions that could somehow affect the extractivist bias embedded in the computation of earth's depths and densities, and the ways in which this organizes life.

As the so-called earth spun before us, the universalist geologic commons emerged.[12] A particular regime embedded within the software that imbues the *histories of colonial earth-writing* and a *geologics* in which "[e]xtractable matter must be both passive (awaiting extraction and possessing of properties) and able to be activated

through the mastery of white men".[13] In these scenes of turbocapitalism, the making present of fossil fuels and metals as waiting for
extraction heavily depend on software tools, such as Gplates, for handling, interpreting and 3D visualization of geological data. These
entangled softwares form an infrastructural complex of mining and
measuring. Such tools belong to what we refer to as "the contemporary regime of volumetrics", meaning the enviro-socio-technical politics — a computational aesthetics — that emerge with the measurement of volumes and generation of 3D objects. A regime full of bugs.

Reporting a bug, bugging a report

*Somewhere there is a fault. Sometime the fault will be
activated. Now or next year, sooner or later, by design, by
hack, or by onslaught of complexity. It doesn't matter. One
day someone will install ten new lines of assembler code,
and it will all come down.*[14]

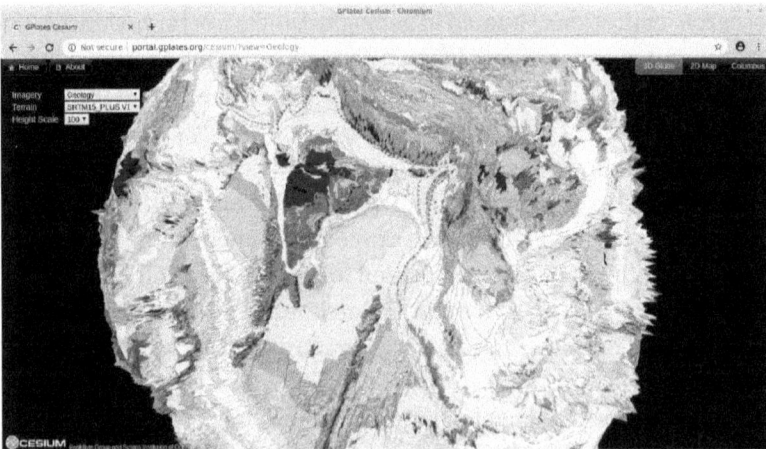

Gplates web portal: Geology view. Earthbyte Group and Scripps Institution of
Oceanography, accessed June 1, 2019, https://portal.gplates.org/

Bug reporting, the practice of submitting an account of errors, flaws,
and failures in software, proposes ways to be involved with technological development that not only tolerates, but necessarily requires
other modes of expertise than writing code. Bug reporting is a lively
technocultural practice that has come to flourish within free software
communities, where Linus' law "with many eyeballs, all bugs are shallow" still rules.[15] The practice is based on the idea that by distributing

the testing and reporting of errors over as many eyes (hands, screens, and machines) as possible, complex software problems can be fragmented into ever smaller ones. By asking users to communicate their experiences of software breakdowns effectively, bug reporting forces "the making of problems" through a process of questions and fragmentation.[16] It exposes so-called bugs to a step-by-step temporality, to make even the hardest problems small enough to be squeezable,[17] as they eventually are reduced to nothing more than *tiny* bugs.

In order to streamline the process of such squeezing, many software platforms have been developed to optimize the cycle of bug reporting and bug fixing.[18] "Issue trackers" help developers to separate bug reports from feature requests. A "bug" is a fault or an error that responds to what is already there; a "feature request", on the other hand, is a proposal that adds to the project-as-is; it extends an existing feature or ultimately necessitates the rethinking of a software's orientation. It is obvious that in such a technosolutionist framework, reports will attract attention first, while requests have a lower priority. Once identified as such, a bug can then be tagged as "critical" (or not), assigned to a specific piece of code, a software release, a milestone, a timeline, or a developer who then will need to decide whether it is a syntax, run-time or semantic error. From then on, the bugs' evolution from "reported" to "resolved" will be minutely tracked.

The issue with issue trackers and with bug reporting in general is that these are by definition coercive systems. Issues can only be reported in response to already existing structures and processes, when "something is not working as it was designed to be".[19] But what if something (for example, in this particular case, a geocomputation toolkit) is not designed as it should be? Or even more importantly, what if geocomputation should not be designed, or it should be actively undesigned and not exist at all? Or what if there were no way to decide or define, in advance, how something should be without making an authoritative gesture of prejudgment and imposition?

Bug reporting tightly ties users' practices to the practice of development, making present the *relations* of software — it is a mode of practicing-with. Like Haraway's situated practice of writing, figured by Maria Puig de la Bellacasa as a "thinking-with" and "dissenting-within", bug reporting makes apparent that software does not come without its world.[20] Dissenting-within figures as both an embedded mode of practice, or speaking from within open-source software, problematizing an idea of a critical distance; but also has an "openness to the effects we might produce with critiques to worlds we would rather not endorse".[21]

Maybe it is time to file a bug report on bug reporting. Both writing and reading bugs implies a huge amount of empathy, but this is in fact a technically constrained sort of empathy: through steps, summaries, evidences, and indexing the reporter needs to manage her urgency and sync it with that of the wider apparatus of the software's techno-ecology and its concrete manipulator or interlocutor. What if we would use these processes for collectively imagining software otherwise, beyond the boundaries that are drawn by limiting the imagination of what counts as a bug, such as the productivist hierarchization between "features" and "bugs"?[22] Bug reports could allow space for other narratives and imaginations of what is the matter with software, *re-mediating* it with and through its troubles, turning it inside-out, affecting it and becoming affected by it in different ways.[23] "GPlates 2.1 was released today! Many bugs have been fixed, including the computation of crustal thinning factors."[24]

In our attempt to imagine a bug report on Gplates, many questions started to emerge, not only in relation to how to report, but also because we were wondering *whom* to report to. In other words: a repoliticization of the practice of bug reporting implies thinking about the constellation of interlocutions that this culture of filing inserts its sensibilities in. If we consider software to be part of an industrial continuum, subjected to a set of values that link optimization, efficiency, and development to proficiency, affordability, and productive resilience, then where should we report the bug of such an amalgam of turbocapitalist forces? To whom should we submit reports on patriarcocolonialism? It also became clearer that making issues smaller, and shallow enough to be squeezed, was the opposite of the movement we needed to make; the trust in the essential modularity of issues was keeping problems in place.

GPlates for example, confirms users' understanding of the earth as a surveyable object that can be spun, rendered, grabbed, and animated; an object to be manipulated and *used*. There is, as Yusoff notes, no separation between technoscientific disciplines and the stories they produce, but rather an axis of power that organizes them.[25] Gplates is very much part of this axis, by coercing certain representational options of earth itself. But it also does so through computational choices on the level of programming and infrastructure, through interface decisions and through the way it implements the language of control on multiple levels. These choices are not surprising, they align with other geocomputation tools, other volumetric rendering tools, and with normative understandings of the agency and representations of the earth in general.

Could we imagine filing a bug report on Gplates' timeline implementation, insisting on the obscenely anthropocentric faultiness of the smooth slider that is moving across mega-annums of geological time? How would we isolate this issue, and say exactly what is wrong? And since a bug report requires reproducibility, how would we ask a developers' collective to reproduce the issue one more time in order to rigorously study options for nonreproducibility in the future, and what do we expect the collective to do about it? We need a cross-platform, intersoftware, intracommunity, transgenealogical way of reporting that, instead of making bugs smaller, scales them up in time and space and that can merge untested displacements and intersections into its versioning ladder.

The practices of bug reporting could be considered as ways to develop trans*feminist commitments to the notion of thinking-with.[26] This is a mode of engagement with technological objects that is potentially porous to nontechnical contributions; that is: to those by queers, women, people of color, non-adult and other less-entitled contributors. This also means that what seems (and is felt) to be the problem with technosciences has the potential to be arranged in other ways at the site of the bug report. Such porosity for *calibration-otherwise* and in differing domains, opens up through the intense squeezing, fragmentation, and proliferation of problems.[27] This exterminating, almost necropolitical motion of squeezing operates on bugs that are small enough to be killed.[28] Squeezing to kill has as a rough consequence that those who are involved in the killing need to assume the responsibility for considering how and why to force through different conditions for the possible, but not others. Such considerations might generate semiotic and material circumstances for making interventions into the damages that are caused by the practices of geocomputation and software like GPlates.[29] It might be a way to do what we call *queering the damage*, and to extend queer theories concerned with personal injury into geocomputational ensembles in order to consider the effects of damages shared by humans and nonhumans. By practicing *queering damage* in relation to geo-computation, we engage with the injuries caused by these volumetric practices. This is a kind of trans∗feminist practice that does not seek to erase histories of injury and harm, but which recognizes that there is a generative force within injury.[30] A force that might take the form of partial reparations, response-ability, (techno)composting and reflourishing.

While we would like to consider bug reporting as a form of response-ability taking, there is also another option.[31] Instead of staying with the established manners dependent on the existing and hegemonically universalizing logic repertoire for technical processing, we might refuse to fix many tiny bugs under the guise of agile patching and instead consider opening a "BUT" gate.[32] This is a political operation: instead of trying to "fix" the Gplates timeline, we could decide to creatively use it by for example setting the software's default for "present" to a noncorresponding year, or by mentally adding a 0 to each of the displayed numbers. Another way to stay with the trouble of software might be to use things as they are, and to invent different modes by the very practice of persistent use.[33]

Disobedient action-research as a form of bug reporting on research itself

They look over at the group of well-known companions and just-known participants, and ask: "if multiple timescales are sedimented in contemporary software environments used by geophysics, can fossil fuel extractivist practices be understood as time-traveling practices?" They observe that this will need to be a question for the bug report. Running the mouse across the screen turning the software of geophysics, they ponder how, through visualizing plates in particular ways on a timeline, Gplates renders a terra nullus, an emptied world.[34]

This essay started as collective bug report on Gplates software, but in order to file such a report, it needed to disobey the axiom of problem reduction, and zoom out to *report on bug reporting* as a practice. Let's now bug the way research engages itself with the world, and specifically how it affects and is affected by computational processes.

Orthodox modes of producing knowledge are ethically, ontologically, and epistemologically dependent on their path from and towards universalist enlightenment; the process is to answer questions, separate them from each other, and eventually fix the world, technically. This violent and homogenizing attitude stands in the way of a practice that, first of all, needs to attend to the re-articulation and relocation of what must be accounted for, perhaps just by proliferating issues, demands, requests, complaints, entanglements, and/or questions.[35]

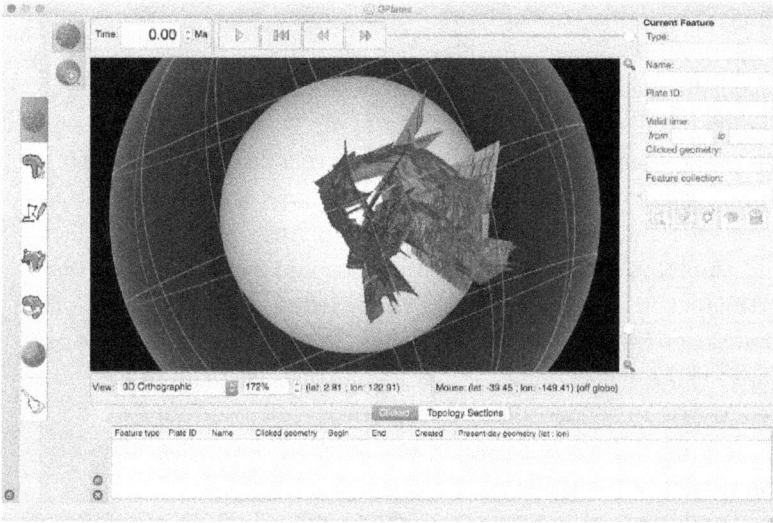

Gplates main interface with data loaded

Take vocabularies as a vector, for example: in order to report on the bug of using the term "grabbing" in Gplates — of which a participant in the *Depths and Densities* workshop astutely observed that "if all the semantic network of Gplates is based on handling and grabbing as key gestures in relation to the body of earth, a loss of agency and extractivist assumption slips in too smoothly, and too fast" — we are in need of research methods that involve direct action and immediate affection into/by the objects of study.[36] She continued: "Also, the use of the verb 'to grab' brings with it the history and practice of 'land grabbing,' land abuse, and arbitrary actions of ownership and appropriation, which has been correlated both with dispossession by the taking of land, and environmental damage." In other words: if orthodox research methods deal with either hypothesis based on observations that are then articulated with the help of deduction or induction, we are in need of methods that affect and are affected by their very materialities, including their own semantics.[37]

It is appealing to consider the practices of bug reporting as a way to inhabit research. As a research method, it can be understood as a repoliticization and cross-pollination of one of the key traditional pillars of scientific knowledge production: the publishing of findings. In this sense, bug reporting is, like scientific research, concerned with a double-sense of "making public": first, it makes errors, malfunctions, lacks, or knots legible; second, it reproduces a culture of a public interest in actively taking-part in contemporary technosciences. The Underground Division considers bug reporting as a way to engage in disobedient action research. By practicing bug reporting, we might anchor our discussions in encounters with the world and the world that composes them — and this is closely related to the practice of *queering damage*.[38] In this way, bug reporting becomes inseparable from the relations it composes with volumetrics, both with the technical and through its relations with queer and feminist theory. Disobedient action research "invokes and invites further remediations that can go from the academic paper to the bug report, from the narrative to the diagrammatic and from tool mis-use to interface redesign to the dance-floor. It provides us with inscriptions, descriptions, prescriptions and reinterpretations of a vocabulary that is developing all along."[39]

Action research as an established method is by definition hands-on, site-specific and directly interpelling systems, and in that sense, it is already close to the potential of bug reporting as a form of response-able research. In a way, action research is always already disobedient, because it refuses to stand back or to understand itself as separate from the world it is researching; with Karen Barad we could say that action research assumes it is "always-already entangled".[40]

The "disobedient" in disobedient action research means it refuses to follow the imagined looped cycle of the evolving timeline of theory and practice. It does not fit the neatly posed questions of a technical bug report neither. It instead works diffractively across the *deep implicancies* of collective research with software, cutting between various lines of inquiry.[41]

The specific disobedience that Possible Bodies brings to Gplates is the refusal to scope according to the probable axis of universalism, productivism, and determinism. It is a disobedience that instead moves across vectors, coordinates, and intersectional scales and — why not? — emerges from within those very vectors and their circumstances. It proposes a calibration-otherwise for volumetrics that can be understood as a form of disciplinary disobedience, a gesture that does not reject scale and the expertise of geocomputation but that problematizes its aftermath while experimenting with other applications and implications.

This disobedient bug report on Gplates therefore needed to ask about temporalities and their material and semiotic conditions, but at the same time concretely wonders how the software imagines the end of time(s), in a modelling sense.[42] Within such diffractive cycles, the disobedient bug report attunes to all types of bugginess within a process: "[the underground] is no longer (or never was) the exclusive realm of technocrats or geophysics experts".[43]

Tuning in to these various lines, disobedient action research has its own liveliness, searching out the bugginess in all tools, forcing a debugging of more than just software, and asking users and developers to consider a commitment to the deep implicancies of earth sciences, extractivism, software development, and coercive naming, to name only a few possible agential cuts. The point of disobedient action research is that the feminist commitment to stay with the trouble is made operational within the work itself.

These ongoing buggy moments of research and reporting then need to include the bugs within Eurocentric, identitarian white feminist theoretical frameworks and practices that we are uncomfortably infused by. The worlds which they are rendering visible worry us, and the ones excluded from this rendering urge us to try harder. As object and subject co/mingle in the bug report, worlds become recast, "where poetic renderings start to (re)generate (just) social imaginations".[44] In taking up the software tools of geophysics research and industry, we are reminded collectively that technical knowledge is not the only knowledge suitable for addressing the situations we find ourselves in.[45] As we anchor our disobedience in trans*feminist figurations, bugs obviously appear in how "we make it otherwise".[46] Rendering through figures, some of our anchors become lost and others become necessarily unstable, as they make certain worlds tangible, and render others absent.

Nonfixing as experimental de/rebugging @ Gplates

What if the earth grabs back?[47]

The attempt to write a bug report on Gplates forces us to reconsider the implications of a fix and its variations, such as the technofix and the reparation. As necessary as it seems to report the damaging concoction of representations, computations, vocabularies, and renderings, it seems important to not assume these issues to be addressed in order to (just) fix them in the sense of putting them back in circulation. Or to say it differently: to change it all so nothing really changes.

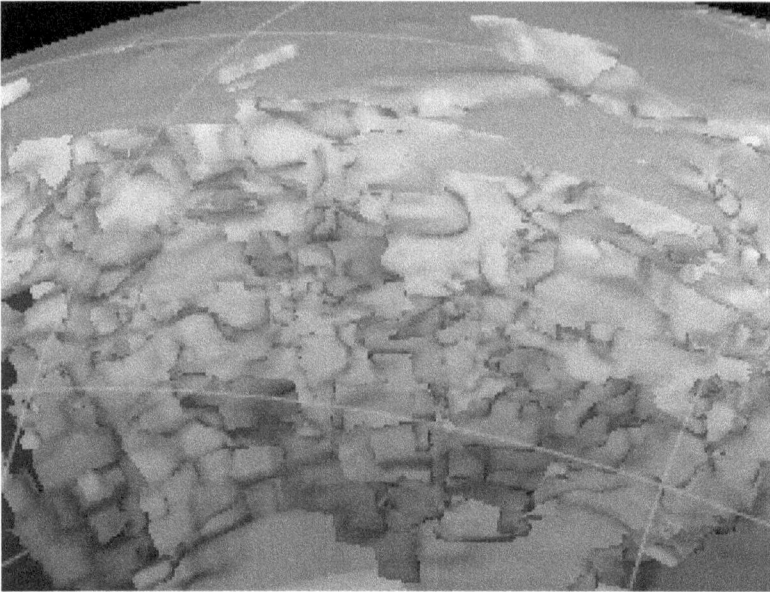

Gplates main interface (detail): grabbing the earth

In the *turbocapitalist momentum*, are there other options besides abrupt deceleration and hyperlubricated acceleration? Are there ways of working without guarantees or attempting to resist ever-new reparative fantasies of technoscience? However, we are not calling for an anti-affirmative stance; but instead by making the leap in scale, together with queer and antiracist ontologies in our software critique we place an emphasis on damages across the industrial continuum of volumetrics. As Heather Love notes, queer practice "exists in a state of tension with a related and contrary tendency — the need to resist damage and to affirm queer existence".[48] In a mode of queering-damage-as-queer-existence, we extend the possibility of intervention from body politics to geopolitics.

To engage together in disobedient bug reporting might be a queer way to learn more sophisticated ways of identifying how regimes of truth, ideology, or representation affect our most immediate and mundane naturecultures. The hegemonic acceleration of contemporary technologies imposes a series of conditions that lead to the persistence of cultural forms of *totalitarian innovation* which must be resisted and contested. Yet those same conditions also constitute a complex of latencies and absences with which we have to inventively coexist, driven by the need for attentive, politicized presences. In a way, the persistent practice of finding "bugs" as another possible mode to conduct research tracks the potential to stay with the trouble of software in a responsible, creative way. The bug reporting on GPlates is an affirmative mode of software critique that refuses to organize along the vectors of reparation or resilience, but rather strives to grab back.

In other words, writing disobedient, collective, situated bug reports might be a method of pushing the limits of the probable and expanding the spectrum of the possible. Discussing technological sovereignty and infrastructural self-defense initiatives are good places to start, but those gestures are certainly not enough.[49] The first step is to methodologically identify and affirmatively publish the damages that coercive turbocapitalism inflicts through volumetrics and geocomputation. We need to join forces and write bug reports on these systems in order to technically equip ourselves with partial and localized repair possibilities, while resisting the production of ever-new and naive reparative fantasies.

As a future work, we started to think about what noncoercive computing would involve, as it becomes increasingly clear that the hubris of, let's say, the Gplates timeline is rooted in the colonial *computationalism* of such a project.[50] It won't all happen tomorrow, but we can start with a rough outline together.

We have always been geohackers.

Notes

1. ↑ Jara Rocha, "Depths and Densities: A bugged report," in this book.

2. ↑ "Depth and Densities: A Possible Bodies Workshop," accessed July 1, 2019, https://2019.pastwebsites.transmediale.de/content/depths-and-densities-a-possible-bodies-workshop.

3. ↑ See also the introduction to this book.

4. ↑ "Gplates, desktop software for the interactive visualisation of plate-tectonics," accessed July 1, 2019, https://www.gplates.org/.

5. ↑ Yusoff, *A Billion Black Anthropocenes or None,* 7.

6. ↑ "Depths and Densities" workshop materials can be found at https://pad.constantvzw.org/p/possiblebodies.depthsanddensities.

7. ↑ "Item 074: The Continuum," *The Possible Bodies Inventory,* 2017.

8. ↑ "GPlates 2.0 software and data sets," accessed July 1, 2019, https://www.earthbyte.org/gplates-2-0-software-and-data-sets.

9. ↑ "Gplates, desktop software for the interactive visualisation of plate-tectonics,

10. ↑ "EarthByte: People," accessed July 1, 2019, https://www.earthbyte.org/people/.

11. ↑ "grey means there is nothing such as a body of earth / it is almost a void / whole parts of grey earth / like you are making a cake / you can put toppings on." In Rocha, "Depths and Densities".

12. ↑ Yusoff, *A Billion Black Anthropocenes or None,* 2.

13. ↑ Yusoff, *A Billion Black Anthropocenes or None,* 2, 14.

14. ↑ Ellen Ullman, *Close to the Machine: Technophilia and Its Discontents* (New York; Picador, 2012), 32

15. ↑ Eric Steven Raymond, "The Cathedral and the Bazaar," accessed July 1, 2019, http://www.catb.org/~esr/writings/cathedral-bazaar/cathedral-bazaar/index.html.

16. ↑ As Simondon notes, "living is itself the generation of and engagement with problems." Gilbert Simondon, *L'Individuation à la lumière des notions de forme et d'information* (Grenoble: Editions Jérôme Millon, 2013).

17. ↑ In the context of technical bug reporting, squeezing refers to fixing.

18. ↑ Issue trackers are increasingly being integrated into software versioning tools such as git, following the increasingly agile understanding of software development.

19. ↑ "Bug: Definition — What Does Bug Mean?," accessed July 1, 2019, https://www.techopedia.com/definition/3758/bug.

20. ↑ See Maria Puig de la Bellacasa, "Nothing Comes without Its World: Thinking with Care," *The Sociological Review* 60, no. 2 (2012): 197-216; Donna J. Haraway *Staying with the Trouble: Making Kin in the Chthulucene* (Durham: Duke University Press, 2016); and Kathrin Thiele's chapter in this publication.

21. ↑ de la Bellacasa, "Nothing Comes without Its World: Thinking with Care," 205-206.

22. ↑ "Experiments in virtuality—explorations of possible trans∗formations—are integral to each and every (ongoing) be(coming)." Karen Barad, "Transmaterialities: trans∗/Matter/Realities and Queer Political Imaginings," *GLQ: A Journal of Lesbian and Gay Studies* 21, no. 2-3 (2015), 387-422.

23. ↑ "I find bug reports interesting because if they're good, they mix observation and narration, which asks a lot from the imagination of

both the writer and the reader of the report." Femke Snelting and Christoph Haag, "Just Ask and That Will Be That (Interview with Asheesh Laroia)," in *I Think That Conversations Are the Best, Biggest Thing that Free Software Has to Offer*, (Brussels: Constant, 2014), 201-208.

24. ↑ "GPlates 2.1 released (and pyGPlates revision 18)," accessed July 1, 2019, https://www.earthbyte.org/g plates-2-1-released-and-pygplates-revision-18/.

25. ↑ "There is not geology on one hand and stories about geology on the other; rather, there is an axis of power and performance that meets within these geologic objects and the narratives they tell about the human story. Traveling back and forth through materiality and narrative, the origins of the Anthropocene are intensely political in how they draw the world of the present into being and give shape and race to its world-making subjects." Yusoff, *A Billion Black Anthropocenes or None*, 34.

26. ↑ de la Bellacasa, "Nothing Comes without Its World: Thinking with Care," 199.

27. ↑ For a discussion on recalibrating relations, see Helen Pritchard, Jennifer Gabrys, and Lara Houston, "Re-calibrating DIY: Testing digital participation across dust sensors, fry pans and environmental pollution," New Media & Society 20, no. 12 (December 2018): 4533–52.

28. ↑ Joseph-Achille Mbembe and Libby Meintjes, "Necropolitics," *Public Culture*, 15, no. 1, (2003), 11-40.

29. ↑ Although there is not enough room to expand on the damages here, they include pain, suffering, injury, uselessness, homophobia, racism, ageism, ableims, specism, classism, exclusions, and inclusions.

30. ↑ For more context on queering damage and extending queer theories of injury see "Queering Damage," accessed July 1, 2019, http s://queeringdamage.hangar.org; and Helen Pritchard, *The Animal Hacker*, Queen Mary University of London, 2018, 244.

31. ↑ "Blaming Capitalism, Imperialism, Neoliberalism, Modernization, or some other 'not us' for ongoing destruction webbed with human numbers will not work either. These issues demand difficult, unrelenting work; but they also demand joy, play, and response-ability to engage with unexpected others." Donna Haraway, "Anthropocene, Capitalocene, Plantationocene, Chthulucene: Making Kin," *Environmental Humanities* 6, no. 1 (2015): 159-165.

32. ↑ A "BUT gate" is a proposed addition to the logic operators at the basis of electronics and computation, a gate that would halt the process and make time to discuss concerns on other levels of complexity. More about this proposal by ginger coons and Relearn in "Item 013: BUT: an additional logical gate," *The Possible Bodies Inventory*, 2014.

33. ↑ Donna Haraway, *Staying with the Trouble*, 2.

34. ↑ A recounted scene from the "Depths and Densities" workshop. On terra nullius, see Marisol de la Cadena and Mario Blaser (eds.) *A World of Many Worlds* (London, Duke University Press, 2018), 3: "The practice of terra nullius: it actively creates space for the tangible expansion of the one world by rendering empty the places it occupies and making absent the worlds that make those places."

35. ↑ "Without separability, sequentiality (Hegel's ontoepistemological pillar) can no

longer account for the many ways in which humans exist in the world, because self-determination has a very limited region (spacetime) for its operation. When nonlocality guides our imaging of the universe, difference is not a manifestation of an unresolvable estrangement, but the expression of an elementary entanglement." Denise Ferreira da Silva, "On difference without separability" i*32a São Paulo Art Biennial catalogue: Incerteza viva/Living Uncertainty* (São Paulo: Publisher, 2016) 57-65.

36. ↑ Rocha, "Depths and Densities: A bugged report," in this chapter.

37. ↑ de la Bellacasa, "Nothing Comes without Its World: Thinking with Care," 199.

38. ↑ Marilyn Strathern, "Opening Up Relations," in *A World of Many Worlds*, Marisol de la Cadena and Mario Blaser, eds. (London: Duke University Press 2018).

39. ↑ Jara Rocha, and Femke Snelting, "Dis-orientation and Its Aftermath," in this book.

40. ↑ Karen Barad, *Meeting the Universe Half Way* (Durham: Duke University Press, 2007).

41. ↑ Denise Ferreira da Silva, and Arjuna Neuman, *4 Waters: Deep Implicancy* (2018).

42. ↑ Rocha, "Depths and Densities: A bugged report."

43. ↑ Rocha, "Depths and Densities: A bugged report."

44. ↑ Possible Bodies (Helen V. Pritchard, Jara Rocha, Femke Snelting), "Ultrasonic Dreams of Aclinical Renderings," in this book.

45. ↑ Lucy Suchman, *Human-Machine Reconfigurations: Plans and Situated Actions: Second Edition* (Cambridge: Cambridge University Press, 2006), 188-190.

46. ↑ Elizabeth A. Povinelli, "After the Last Man: Images and Ethics of Becoming Otherwise," *e-flux* 35 (2012).

47. ↑ Rocha, "Depths and Densities: A bugged report."

48. ↑ Heather Love, *Feeling Backward* (Cambridge MA: Harvard University Press, 2009), 3, cited in Helen Pritchard, *The Animal Hacker*, PhD thesis, Queen Mary, University of London, 2018, 244.

49. ↑ Alex Haché, et al. eds., Soberanía Tecnológica (Ritimio, 2014), https://www.plateforme-echange.org/IMG/pdf/dossier-st-cast-2014-06-30.pdf.

50. ↑ Syed Mustafa ali, "A Brief Introduction to Decolonial Computing," *XRDS: Crossroads, The ACMMagazine for Students* 22, no. 4 (2016) :16-21.

LiDAR on the Rocks

The Underground Division
(Helen V. Pritchard, Jara Rocha, Femke Snelting)

LiDAR on the Rocks is a training-module for hands-on collective investigation into the micro, meso and macro political consequences of earth scanning practices. The module looks into how undergrounds are rendered when using techniques such as Terrestrial Light Detection and Ranging (LiDAR), magnetic resonance, UltraSound, and Computer Tomography (CT).

Preferably when surrounded by fake rocks, use green string and yellow tape to manually construct point clouds and experiment with Point of View (POV). Try to render intersecting positions and shift from individual to collective pareidolia (seeing worlds inside other worlds), while reading selected text fragments by N.K. Jemesin,[1] Kathryn Yusoff,[2] Elizabeth Povinelli,[3] Karen Barad[4] and Denise Fereira da Silva.[5] Each session ends near a $1m^3$ area of grass that is marked for imagined digging, plus a DIWO metal detector to provoke plural renderings of the underground.

Participants in *LiDAR on the Rocks* can now be introduced into the Initial Areas of Study (IAS) of *The Extended Trans ∗feminist Rendering Programme* (T∗fRP):[6]

- connected subsurfaces
- stories of the undergrounds (sub-terranean science-fiction)
- subsurface politics and its constellations

Notes

1. ↑ N. K. Jemesin, *The Fifth Season (The Broken Earth Trilogy #1)* (Orbit, 2014).

2. ↑ Kathryn Yusoff, *A Billion Black Anthropocenes or None* (Minneapolis: University of Minnesota Press, 2018).

3. ↑ Elizabeth A. Povinelli, "Can rocks die?" *Geontologies: A Requiem to Late Liberalism* (London: Duke University Press, 2016), 8-9.

4. ↑ Karen Barad, "TransMaterialities: trans∗/Matter/Realities and Queer Political Imaginings," *GLQ* 1, no. 21 (2-3) (June 2015): 387–422.

5. ↑ Denise Ferreira da Silva, "In the Raw," *e-flux Journal* #93 (September 2018), https://www.e-flux.com/journal/93/215795/in-the-raw/.

6. ↑ "The T∗fRP exists to take care of the production, reproduction and interpretation of DIWO scanning devices and scanning practices within the field of a-clinical, underground and cosmic imaging. The programme invites fiction writers, earth techno-scientists and trans∗feminist device problematizers to render imaginations of the (under)grounds and of the earth." The Underground Division, "The Extended Trans∗feminist Rendering Programme," 2019, https://possiblebodies.constantvzw.org/rendering/transfeminist_rendering_prospectus.pdf.

Ultrasonic Dreams of Aclinical Renderings

Possible Bodies
(Helen V. Pritchard, Jara Rocha, Femke Snelting)

Note: When reading, make sure to listen to the soundtrack provided here: http://volumetricregimes.xyz/files/MRI_SOUNDINGS.mp3

1. Evening

The machine began to rotate slowly. She swallowed the paramagnetic contrast agent in one go, preparing her vessels to render themselves later. When the metallic taste faded, she could smell the ancient chestnut trees blossom nearby. Her crystal studded belt was stored with the pyrosome pendant in a strongbox outside the perimeter and the radio-pharmaceutical body-paint shimmered, still wet. Across from her, the others followed and struck an A-pose. Judging by the roar of the crowd that was barely audible from inside, tonight they would finally make a living.

Following their post-certification dreams, they ran their own techno-ecological show in excess of vision. The machine was rigged together from a salvaged General Electric Discovery MR750w and a Philips Ingenia 3.0T. For effect, several pieces from a scanner built in the seventies by the Electric and Musical Industries conglomerate had been added. This aclinical setup had cost virtually a million but when dismantled, the hardware fit on a standard trailer and the open sourced software did not take up more than two solid-state drives. The certificates doubled as a license for speculative imaging and now their only worry was how to pay for the astronomic electricity bills without starting a forest fire.

The lights dimmed and the noise grew louder until all solids vibrated: bones, glass, teeth, screws, violently rattling. They squeezed each other tightly as the machine picked up pace, centrifugal forces flattened their bodies against the curved superconductive screen behind. The ground dropped away and an electromagnetic coil lit up in the centre.

Now they all moved together, more-than-human components and machines, experiencing an odd sensation of weightlessness and heaviness at the same time. Limbs stuck to the wall, atoms bristled. Bodies first lost their orientation and then their boundaries, melting into the fast turning tube. Radiating beams fanned out from the middle, slicing through matter and radically transforming it with increasing intensity as the strength of circlusion decreased. The sound of the motors became deafening when the symmetric potential excited the rotating matter, pulling the cross-sectional spin-spin couples towards the central coil, forcing atomic spectra to emit their hyperfine structure. Once all fluids were accounted for, the volumes could be discretely reduced to graphs and the projections added up. Attenuating varying levels of opacity, a white helix formed in the middle which slowly gathered intensity and contrast. Faster and faster the machine spun until the cylindric screen lit up in the dark.

When the shadowgraphs appeared, the crowd howled as coyotes. Laminograms of differently densed matters rendered onto and through each other, projecting iteratively reconstructed insides onto the outer surface area. Collarbones entangled with vascular systems. Colons encircled spinal chords and a caudal fin, a pair of salivary glands vibrated with a purring larynx at a frequency of 25 to 150 Hertz. Brain activity sparked cerebral hemispheres, creating free-floating colonial tunicates of pulmonary arteries mingling with those of lower legs.

The math was breathtaking. Volumetric figures pulsated back and forth between two to three dimensions, transforming images into accidented surfaces and surfaces into ghostly images. There were mountain areas divided by sharp ridges, and watersheds preventing the draining of enclosed reservoirs. Methane leaked out of the old wells below and caused tiny explosions each time an image hit the surface. Calculating the distance between the edges of those catchment basins, the exponential boundaries between objects computed on the fly. There were dazzling colors as the sinographs peaked and the cubes marched. Whirling polygonal meshes exploded into a cloud of voxels before resurfacing as new nauseating contours, trapped in the vapours of the display. The continuing presence of the leftover, remnant of the former plutonium plant mightpotentially include anything that had escaped the nature refuge.

2. Night studies

- Hey more-than-human components and machines, how are you?
- Let's meet every night at the school party! We will silently split up and follow our ears.

- From now on, the learning happens at that precise moment when the co-participating spectrum produces a kind of blue that emerges up to 90 feet (30 m) in clear water. How will that sound?
- At night we persistently learn to sense the emitted reflected radiation remotely, as a tactic for profanating the image-life industrial continuum.
- We will gather to body-image geological structures, heat differences in water currents. We'll also otherwise embody others, and start fires – a significant activity these days, you know.
- Let's make sure to reserve our electric sockets, before the curricula sediments. Some of us might highlight the urge to involve many more not-only-human companions, just like ourselves.
- Whoops! Over there others claim that all of this is happening precisely thanks to how non-supervision has already functioned quite accurately for eons; everybody will perhaps nod and we will start computing together.
- Key to our program is that the n-dimensionality of unsupervised machine learning radicalizes the project to the nth power.
- Each learning machine decrypts a split for their teaching fee, a fraction of the full amount that we spend on whatever desire, a software fantasy or whatever we want. Or for cigarettes.
- The one condition will be that we commit to talk about what to do with the tokens, and how to calculate the coins. In our meetings this is such a frequent consensual mode. At other times, glossy dissent might take place.

This is how it goes:

At first we are buried and cemented in, and we cannot get through. But then a flower breaks through the asphalt and the old regime of waves is finally over. A radical symmetry of processing agencies materializes. There is no evaluation any more: this is the take of the spectrum. Despite the cost of electricity and the heat from the rapid fires, now we just can't get enough.

The four dimensions of our learning program are: depth (z), height (y), width (x) and time (t). Although some have argued for the dimension of affect (a), it is settled this is always already present here or, to put it differently, affective dimension is always-already inter-sectional. The program is open and rigorous:

z) For deep structures of either objectification or subjectification, or both, or third parts, in z they train "profound imaging". We learn to estimate our present density without classifying it.

y) The principle of the inverse problem: "While the object or phenomenon of interest (the state) may not be directly measured, there exists some other variable that can be detected and measured (the observation) which may be related to the object of interest through a calculation." Exercising this problem can lead to an inversion into a stateless level. This is technically understood as "low profiling".

x) Crystalogy it is. Gymnastic practice for the expansion of chosen prismatic geometric splendours.

t) Frequency. This module goes into the ontologies of ongoingness. In-determinate waving. An intensive training to not be always available.

The four dimensions are rendered through continuous intra-actions with various devices and techniques. Machinic learners that are supposed to experiment with and be experimented on include (but are not limited to): computer tomography, magnetic resonance imaging and ultra-sound. While the frequency is mandatory, tech-niques, physicians, bodies are requested to certify each other intra-actively.

The schedule is almost full. Mid-red produces the worlding of vegetation, soil moisture content and, in some cases, forest phenom-ena. A heavy piezoelectric glow emits from the zone where sensitive detectors are placed. The tuners are humming, tuning with frequent errors. Neither the production nor the interpretation of ultrasound images are simple matter; mis-diagnosing mis-readings involves highly specialized forms of knowledge.

The party is going on. 'The spectrum is no longer (or never was) the exclusive realm of technocrats or medical experts', says a banner on the wall. That bunch of new wave spectrometers, speedy spec-trophotometers, cats, or dark industrial spectral analyzers are shak-ing and hot. Turning around into something else. Our in-determinate ontologies are here to stay... or maybe not. With care, curiosity and passion, dissonant matters are all being made present. There is no discriminatory weight, but for sure there are mutual exclusions that need to be accounted for. Here subaltern scopes are critical and (still) celebrated. We are considered as rich, exuberant and glossy in our fierce but so-called-precariousness. From now on, language will need to inflate and mutate to fit the hyperspectral sensing; reading lists are not printed here. Until we reach the no-mattering-morning, we still have many nights to spend responsibly, living ourselves collectively in an exuberant way. A shy crew in an immanent shiny excess. Hell yeah.

When the light changes again, we finally finish. It works as a signal to shoot. We are exhausted but once propagated, our unlearned signals keep training on their own: unsupervising others, reversing geometries, undetermining yet-to-know subject-object mining. Our dreaming vigilance is the same at 9 am as at 2 am.

From now on, hyperspectral imaging takes advantage of the spatial relationships among the different spectra in this specific neighborhood of blurry limits. It is placed in practice to generate more elaborate misreadings of spectral-spatial accuracy models for the subsequent segmentation and classification of the image (otherwise understood as imagination). Sheer volume.

Check out that very corner, how it shows its complex composition. The low frequency but high-res flickering. Filled with noisy false colors.

Check out that roof over there, its densities deserve to be seen. Those sexy hyperspectrals are being rendered continuously. Let's follow them all the way into ultrasonic cosmo-dreaming.

Here-now. It is finally the moment to take the means for themselves. Everyone is here. The whole spectrum is present, and making itself present.

3. Day 9

Certified, the night studies programmers lay as still as possible. With their hands flat on the damp soil, bodies a faint outline along the edge of the drill site, they prepared for the ninth day computed tomography earth scan. At the night studies they assumed the were now activists. She was still clutching an instruction leaflet that read "Image Wisely Programme – sign on in advance to an adventure that will leave none of the terms we normally use as they were". Under the dusk light the recently rigged solar panels shimmer against the device mirrors. Some of them were soldering connections over the soil with their portable irons, connecting the scanners across the earth's surface to the super computer user. In the reflection of her screen, she could see across the crowd a tangle of wires trailing out to fault lines, and as they draped these wires over their bodies in preparation, a long high pitched drone started to sound – as if a balloon was letting out air. In the distance, the dogs started barking a scene of wilding activities, they had learned about the possibility of this during training. The devices had begun. Infecting the entire structure as a whole. An electric field desiring a field born of charged yearnings. Cell death.

Earth bodies no longer accepting of the role assigned to them were beginning to emerge from the orbiting electrons, a few days and night had past but they seem to have lost count and felt somewhere in between, apart from when the speaker sounded to the Unix time-stamp announcing the day, hour, minute and second of the slice. Dark regions began hitting the photographic film fastened on the back of an old protest banner. The banners were propped up behind them, dark regions beyond expertise. These dark regions were now infected by a different purpose. She shivered, her fur bristled and a layer of cold fell over the crowd. Someone smoking a cigarette draped a leather jacket over her shoulders. It smelled like cattle, tannin and fashion magazine cologne. As they turned, and rotated, an earth-body, they listened into photons, bursting with innumerable imaginings of what might yet(have) be(en). She listened carefully, concentrating for rumors she had not heard before. Densities she had not experienced. Stories set into motion the moment they spill. Addressing intensities.

It was the ninth day of the scan and their bodies began to understand what their ears could not. The difference between a dream and a nightmare – kinetic energy, a net positive electric charge, material wanderings/wonderings began to burst through the earth's surface, sending rays through them. They had discussed this possibility at the training camp. Three dimensional patterns began to divide the absorption of the earth beneath them. A diagnostic system. Water, strata, bone, skin, began to absorb the rays at differing rates. X-rays were traveling outward in some general direction hitting atoms – a quivering electric field. Together they were rendering fractures, internal structures of earth bodies. Here comes the math. Layering slices on top of each other building a three dimensional image. Tissues, microbes, minerals, systems superimposed on top of one fuel capitalism of the past. Beyond any hope of a recuperation but instead searching for the refusals of the mineralized past.

In this picture the voice over the tannoy exclaimed "sacramental plurality". The super computer user was shifting, forming an image of the cross section of the body that could be read on the salvaged screen. Data layering on top of one another to form the entire superuser organism. As the machine body rotated, electrons continued to be produced. Electrons colliding with atoms, transmitting through the entire body these electrons sources. A pleasuring intensity of measurements at all possible partial angles. They were awash with a thickness, a plurality of experiences occurring simultaneously – like a person walking by. Intensities began to break up, the different transition rates, and a voice started to sound numbers. As the final timestamp was called, the gnu begun to gather on the edges of the drill site, occasionally drinking from the runoff pools, with their blunt muzzles and waiting for the signal.

It felt like days before the algorithmic processes wound down, for the machine to slow down and the gravitational pull to take hold again. Slowly intensities were reduced and attenuated. Voxels of bone and mineral started quivering as they were numbered. MR750w. Gradually restricting the handful of variables, the ground came back up and one by one the bodies slid down from the walls that had heated up under the strain of intensive calculations. The high pitched drone stopped sounding and the usher began to take down the barriers. They blinked at each other across the dim radius, faintly glowing, still resonating.

4. Certification

The Extended Trans∗feminist Rendering Program exists to take care of the production, reproduction and interpretation of DIWO scanners and scanning practices within the field of a-clinical imaging such as magnetic resonance (MR), UltraSound (US) and Computer Tomography (CT). Organized around autonomous, ecologically sustainable municipalities it benefits the scanning equipment themselves, as well as the local amateur operators who interact with a-clinical renderings and speculations. For the unsupervised professionals, certification provides possibilities, Optical Character Recognition, the potential for machine recruitment, increased learning power and electricity tokens. For the program participants, prefigurative organizing certification for MR, US and CT. The Program offers its help to readily identify competent scanner mentors in participant communities.

The rendering program is based upon a set of Crystal Variation Standards that undefine what a competent trans*feminist scanner operator could imagine and might be able to do. Upon fulfillment of these standards, applicants are granted the ET*fRP Professional Certification credentials.

Framed within the ET*fRP, learning forks lead to a number of specialized degrees, including:

- Agile 2D to 3D Tu(r)ning.
- Interpretation of Diversity.
- Radiation Safety and Self-Defence.
- Recreational Imaging.
- Cut, slice and go.
- Neolithic Temporality: theory and practice.

Please bring sufficient electricity tokens, bandanna or blindfold, blanket (in case you get cold), and if possible a pillow, to the group meetings. Jewelry and other metal accessories are not allowed for safety reasons. Everything can be a distraction, especially feelings – if you want to cry, you should and use them in the scans and throw a party. You will receive a copy of any one of the following books and cosmology cards by CT1010 of your choosing: Scanner Magic, CT Ceremony, Coyote Spirit Guides (or Pocket Guide to Spirit Machines), Groups and Geometric Analysis: Integral Geometry, Invariant Differential Operators, and Spherical Functions, Choose Your Own Scanning Family, Voxcell Constellations as a Daily Practice, Earth Technomagic Oracle Cards, Cosmic Cat Cards, Messages from Your Cellular Desire Guides, Voxel Algorithm Oracle Cards or Resonating on Gaia at the first meeting. Print on demand.

You must complete each class in sequence!

Bibliography

"Angular Momentum Coupling," Wikipedia entry https://en.wikipedia.org/wiki/Angular_momentum_coupling.

Barad, Karen. "TransMaterialities trans*/Matter/Realities and Queer Political Imaginings." In GLQ: A journal of lesbian and gay studies 21, no. 2-3 (2015): 387-422.

"Breve Gramática de Quechua." Pontificia Universidad Católica del Perú http://facultad.pucp.edu.pe/ciencias-sociales/curso/quechua/gramatica.html.

Bookchin, Murray. "The Social Matrix of Technology." In *The Ecology of Freedom: The Emergence and Dissolution of Hierarchy*. AK Press, 1982.

DrPhysicsA. "CT (Computed Tomography) Scans – A Level Physics." Uploaded May 21, 2012. https://www.youtube.com/watch?v=BmkdAqd5ReY.

electrovlog. "DIY earthquake detector." Uploaded on the electrovlog channel November 27, 2011. https://www.youtube.com/watch?v=hZEtgCwJ7F0.

"General Electric Magnetic Resonance Imaging." (product page), accessed November 27, 2019. http://www3.gehealthcare.com/en/products/categories/magnetic_resonance_imaging

"Hyperspectral Imaging." Wikipedia entry, accessed November 12, 2019. https://en.wikipedia.org/wiki/Angular_momentum_coupling.

Lai, Larissa. *When Fox is a Thousand*. Arsenal Pulp Press, 2004.

Malkoff, Dave. "A CT Scan for Earth." Weather Channel, 2013. https://www.youtube.com/watch?v=TJKYXPlzYI4.

"Philips Healthcare: MRI innovations that matter to you," (product page). Accessed November 27, 2020. https://www.usa.philips.com/healthcare/solutions/magnetic-resonance.

"Rapid Eye, delivering the world, a BlackBridge Planet Labs scanning project." Accessed November 27, 2020. http://web-dev.rapideye.de/rapideye/products/monitoring.htm.

Stengers, Isabelle. *Thinking with Whitehead: A Free and Wild Creation of Concepts*. Translated by Michael Chase. Harvard: Harvard University Press, 2011

Starhawk. "Earth Activist Training." Available at http://starhawk.org/.

Schuppli, Susan. "Radical Contact Prints." In *Camera Atomica*, 277-291. London: Black Dog Publishing, 2015.

University of Antwerp. "Visielab. Computed Tomography and ASTRA Toolbox training course." Uploaded September 10, 2015. http://visielab.uantwerpen.be/computed-tomography-and-astra-toolbox-training-course.

Ward, Kym. "Circluding." in this book.

Weigal, Michael. "The Scanner Story." EMITEL, 1977. https://www.youtube.com/watch?v=dBulN83zjuM.

+ various templates for certification programs.

Soundtrack

Soundtrack collated by Possible Bodies, 2017. Includes fragments from 'BIDE', Diffusion Tensor Imaging, Fluid Attenuation Inversion Recovery, Gradient, K.I.S.S., R.A.G.E., T1, T2., recorded by K. Williams at the Radiology Lab at the University of Iowa Hospital, 2010, http://www.cornwarning.com/xfer/MRI-Sounds.

Appendix

The So-called Lookalike

Dear designer,

My name is Manetta and I am writing you in the middle of a design process and collaboration with Possible Bodies. Together we are turning the *Volumetric Regimes* research into a book. It will be published by Open Humanities Press (OHP) as part of the DATA browser series for which a layout template has been designed a few years ago.[1]

I realize now that I waited as long as I could, but today I could not postpone this part of the process any longer: we need to figure out how to work with this template. The series editors insisted that we could take the design of the book into our own hands, as long as we would "follow the template". They did not specify what this would mean exactly, but made clear that it was important for them that we honored the original design. However, the way the DATA browser books were produced so far is quite different from the way I work, so we need to find a way to produce this layout otherwise, from scratch.

First, a little bit about myself. My design and research practice is shaped by (and shapes)[2] free software, tool making and collective work. This practically translates into layouts being generated by scripts, books being rendered out of webpages and editorial workflows being transformed into collaborative environments. Working in this way allows me to stay with the complexities of technology, learn about the implications of layout software and approach the profession of design as an embedded networked practice.

For Possible Bodies it was important to take design decisions and tools into account as part of the content of the book, and to extend the conditions of openness provided by OHP, by not only making it possible for readers to download the PDF, but to also enable them to learn about and reuse the editing, production and design process itself.[3]

So today the strange game of sticking to our commitment to make this book look *just* like the other books in the series starts. But where to start? The InDesign files that were used for the other books in this series cannot be opened with any of our free software tools. Besides not being able to open these files, also the iterative workflow we have set up to collaborate as a designer with the editors on the design and the editing, does not match.

We're working with a self-hosted MediaWiki platform, a *wiki* in short, that we as editors and designer can use to edit and structure the materials. From this wiki, we download and reformat everything into a single webpage, which becomes the main document that will be styled and turned into a layout using CSS3 paged media standards. We use the Javascript library Paged.js to paginate this layout in the browser and render it as a PDF. The designer and editors both have access to this rendering process, which allows us to approach the editing and design as one continuous process.

Paged.js is one of the tools that are available for making publications using web-to-print technologies. What is special about Paged.js, is their position within the environment that the project is part of and depending on, by being in close conversation with the W3C consortium. The W3C is the international organization for web standards, such as HTML5 and CSS3, that decides which features will be supported by modern browsers and which won't. I'll end there, but the range of people, tools and environments that co-shape this design practice is much larger. The networks of networks of people working with similar attitudes and sensibilities are actually indispensable for making this design practice possible and viable. And who knows what will happen afterwards, once the material[4] and documented code[5] is published online and available for re-use, thanks to the CC4r (Collective Conditions For Reuse) license.[6] So besides the impossibility of trying to link up with a different and (by now for me) quite alien set of tools and ways of working, I am wondering how to reconnect your template and aesthetics to the way this book is being made?

Also the face on the cover of *Volumetric Regimes* is made in a different way as the others in the series. We decided to use *Multi Remix* made by Winnie Soon and Geoff Cox in which the face is not constructed out of typographic characters,[7] but instead made out of variable geometric shapes using the Javascript library p5.js. *Multi Remix* is a code exercise in the textbook *Aesthetic Programming*,[8], where it is contextualized with a note on the face as a static technological symbolic object and "imperial machine": "The face is part of a surface that promotes sameness and ultimately rejects variations."[9]

We found a slab serif font that aligned with the one in your template and we figured out how to replicate the layout's dynamics and overall structuration. But many questions still haunt us: How could this book actually ever pass as a lookalike of the so-called Stuart Bailey template, while the template, as a design device in a general sense, prescribes an agile gesture of implementation for yet another bundle of content? What agency do design-interventions have once a template is set? How to design this book as part of a trans∗feminist responsibility with the world-making praxis that design implies, shoulder to shoulder with the many companions that worked on editing (and hence caring for) this volume?

Of course these questions open up more questions, like, for example, the consideration of situated design, the implications of declutching content from form, the assumptions of optimisation, agility and efficacy as editorial values... and so forth.

I decided to write you a letter as a way to reflect on this strangeness, and hopefully along the way I managed to describe how this book is radically different from the other books in the series. This is not a complaint, quite the contrary. It is a way of making space to imagine different kinds of embedded networked design practices and to better understand how different ways of working are shaped by (and shape) different realities. As much as this is a letter to you, it is most of all a fan-letter to variability, transitioning, situated techno-creativity and multi-centered transformation; an attempt to bring in another kind of (surely crooked, interrupted, knotted and yet to be known) lineage or genealogy of worlding through design practices.

– Manetta Berends

Notes

1. ↑ "Design," *DATA browser series*, accessed October 20, 2021, http://data-browser.net/design.html

2. ↑ Inspired by the motto of the Libre Graphics Research Unit, "TOOLS SHAPE PRACTICE SHAPE TOOLS", https://osp.kitchen/.

3. ↑ "Without wanting to suggest that FLOSS itself produces the conditions for non-hegemonic imaginations, we are convinced that its persistent commitment to transformation can facilitate radical experiments, and trans∗feminist technical prototyping." Possible Bodies, "Volumetric Regimes: Material cultures of quantified presence," in this book.

4. ↑ The Volumetric Regimes wiki is an ongoing workspace for a book in the making and can be found at: https://volumetricregimes.xyz.

5. ↑ The code that has been used to produce this book can be found at: https://git.vvvvvvaria.org/mb/volumetric-regimes-book.

6. ↑ "The authored work released under the CC4r was never yours to begin with. The CC4r considers authorship to be part of a collective cultural effort and rejects authorship as ownership derived from individual genius. This means to recognize that it is situated in social and historical conditions and that there may be reasons to refrain from release and re-use," accessed

October 20, 2021, https://constantvz w.org/wefts/cc4r.en.html.

7. ↑ The generated faces on the covers of the other books in the DATA browser series are taken from the iOS software app Multi, made by David Reinfurt. http://www.o-r-g.co m/apps/multi.

8. ↑ *Multi Remix* appears in the chapter "Variable Geometry," in *Aesthetic Programming*, which is "exploring the technical as well as cultural imaginaries of programming from its insides." Winnie Soon, Geoff Cox, *Aesthetic Programming* (London: Open Humanities Press, 2021), http s://aesthetic-programming.net/.

9. ↑ "In *A Thousand Plateaus*, Gilles Deleuze and Félix Guattari conceive of the face as 'overcoded', imposed upon us universally, resonating with some of the comments we made earlier in this chapter about Unicode. Their main point is that the face — what they called the 'facial machine' — is tied to a specific Western history of ideas (e.g. the face of Jesus Christ). This, in turn, situates the origins of the face with white ethnicity (despite Jesus's birthplace) and what they call 'facialization' (the imposition onto the subject of the face) has been spread by white Europeans, and thus provides a way to understand racial prejudice: 'Racism operates by the determination of degrees of deviance to the White man's face... The face is thus understood as an 'imperial machine', subsuming language and other semiotic systems. The face is part of a surface that promotes sameness and ultimately rejects variations." Soon and Cox, *Aesthetic Programming*.

Publication History

Dis-orientation and its Aftermath — An earlier version of this text was published as: Jara Rocha and Femke Snelting, "The Possible Bodies Inventory: dis-orientation and its aftermath," in *InMaterial*, Vol. 2 Núm. 3 (2017): Cuerpos poliédricos y diseño: Miradas sin límites. A Spanish translation of this text was published as: Jara Rocha y Femke Snelting, "El inventario de Possible Bodies: la des-orientación y sus consecuencias," in *Nmenos1* No. 2, Archivo y procesos del internet (2022).

x, y, z (4 filmstills) — First published as: Possible Bodies, "x, y, z," in *Fictional Journal*, The Uncanny Issue (2018).

Invasive Imagination and its Agential Cuts — An earlier and shorter version of this text was published in Spanish as: Jara Rocha, Femke Snelting, "La imaginación invasiva y sus cortes agenciales," in *Utopía. Revista de Crítica Cultural* (April-June 2019).

The Fragility of Life — First published as: "The Fragility of Life. A conversation between Femke Snelting, Jara Rocha and Simone C Niquille," *Het Nieuwe Instituut Research & Development* (Rotterdam: Het Nieuwe Instituut, 2018).

Somatopologies — Initially created as an installation for Constant_V (Brussels, 2018), *Somatopologies* travelled to the 4th Istanbul Design B iennial; The Exhibition Library, Seoul Mediacity Biennial; LUMA Arles A School of Schools, C-Mine Genk and Goldsmiths, London for the seminar *Volumetric Ecologies.*

The Industrial Continuum of 3D — A Spanish translation of this text will be published in Spanish as: Jara Rocha, Femke Snelting, "El continuum industrial del 3D" (Bilbao: FEM TEK, forthcoming).

MakeHuman — First published as: Jara Rocha, Femke Snelting, "MakeHuman," in *Posthuman Glossary,* eds. Rosi Braidotti and Maria Hlavajova (London: Bloomsbury Academic, 2018).

So-called Plants — Written for the forthcoming publication: *Plants by Numbers,* eds. Helen V. Pritchard and Jane Prophet (London: Bloomsbury Academic, 2023).

Depths and Densities: A bugged report — Report of a workshop with the same title conducted by Possible Bodies feat. Helen V. Pritchard during transmediale 2019. Published as: Jara Rocha, "Depths and Densities: a bugged report," in transmediale journal, issue #3 (Berlin: transmediale, 2019).

We have always been geohackers — First published as: The Underground Division (Helen Pritchard, Jara Rocha, Jara, Femke Snelting), "We Have Always Been Geohackers," in *How to Relate: Knowledges, Arts, Practices* (Bielefeld: Transcript Verlag, 2021).

LiDAR on the Rocks — A first iteration of LiDAR on the Rocks took place at the *Citizen Sci-Fi fair* organized by Furtherfield in Finsbury Park, London on August 10th, 2019.

Ultrasonic Dreams of Aclinical Renderings — First published as: Possible Bodies, "Ultrasonic Dreams of Aclinical Renderings," in *Ada: A Journal of Gender, New Media, and Technology*, No. 13.

Biographies

Manetta Berends works with forms of networked publishing, situated software and collective infrastructures. She is a member of Varia, a member based organisation working on everyday technology in Rotterdam, and an educator at the Masters programme Experimental Publishing at the Piet Zwart Institute.

Sophie Boiron has her graphic designer practice rooted in her photography, painting and cultural management skills. She is participating with Spec uloos studio doing amongst other things book design and typography while gradually expanding the scope of her work as a member of the Atelier cartographique cooperative with focuses on collaborative GIS and paper cartography.

Maria Dada is Lecturer in Interaction Design at London College of Communication. Her work is placed within the fields of design, continental philosophy and visual culture. She investigates the role of digital imagery in reconfiguring socio-political institutions and structures. She has degrees in both continental philosophy from the Centre for Research in European Philosophy and Computing and Communication Arts from the Lebanese American University.

Pierre Huyghebaert is exploring several practices around graphic design, cartography, type design, web interface, schematic illustration, book design and teaching these practices at La Cambre art school. Along with participating in Spec uloos and OSP, he develops topological and non-topological mapping with Atelier cartographique and others Brussels urban projects.

Phil Langley is an architect and "computational designer" based in London. Phil develops critical approaches to technology and software used in architectural practice and more generally for spatial design. As a Director of Bryden Wood Technology, an integrated architectural and engineering practice, Phil leads the Creative Technologies team which is focused on building design automation software for building and infrastructure projects around the world. After training and practising as an Architect, Phil completed his MSc in Architecture:

Computing and Design at UEL in 2007, focusing on generative design and neural networks. He has published and presented his work with software prototypes internationally – in both academic and professional contexts — on the ways in which software mediates design and the built environment.

Nicolas Malevé is a visual artist, computer programmer and data activist, who lives and works between Brussels and London. Nicolas obtained his PhD with a thesis on the algorithms of vision at the London South Bank University in collaboration with The Photographers' Gallery (2021). In this context, he initiated the project "Variations on a Glance" (2015-2018), a series of workshops on the experimental production of computer vision, conducted in several international venues such as Cambridge Digital Humanities Network (Cambridge, United Kingdom), Hangar (Barcelona, Spain), Algolit (Brussels, Belgium), or Aarhus University (Denmark). Nicolas contributed to exhibitions (documenta12, Kassel; Kiasma, Helsinki), research events ("Archive in Motion", University of Oslo; Document, Fiction et Droit, Fine Arts Academy, Brussels; Image Net/Work, Fotomuseum, Winthertur), and publications by MIT Press and Presses Universitaires de Provence.

Romi Ron Morrison is an interdisciplinary artist, researcher, and educator. Their work investigates the personal, political, ideological, and spatial boundaries of race, ethics, and social infrastructure within digital technologies. Using maps, data, sound, performance, and video, their installations center Black Feminist technologies that challenge the demands of an increasingly quantified world — reducing land into property, people into digits, and knowledge into data. https://elegantcollisions.com

Simone C Niquille is a designer and researcher based in Amsterdam, NL. Her practice Technoflesh investigates the representation of identity & the digitisation of biomass in the networked space of appearance. Her work has been exhibited internationally, most recently at HeK-Haus der Elektronischen Künste (2020), Fotomuseum Winterthur (2019), La Gaite Lyrique (2019). She has published writing in *Volume Magazine*, *AD Architecture* and *e-flux*. She is Chief Information Officer at Design Academy Eindhoven. In 2016 she was Research Fellow of Het Nieuwe Instituut Rotterdam and is commissioned contributor to the Dutch Pavilion at the 2018 Venice Architecture Biennale. Niquille is recipient of the Pax Art Award 2020 and Mellon Researcher at the Canadian Center for Architecture (2021/2022). Currently she is investigating the architectural and bodily consequences of computer vision, researching the politics of synthetic training datasets.

Possible Bodies is a collaborative research activated by Jara Rocha and Femke Snelting on the very concrete and at the same time complex and fictional entities that "bodies" are, asking what matter-cultural conditions of possibility render them present. This becomes especially urgent in relation to technologies, infrastructures and techniques of 3D tracking, modelling and scanning. How does cyborgness participate in the presentation and representation of so-called bodies? Intersecting issues of race, gender, class, species, age and ability resurface through these performative as well as representational practices.

Helen V. Pritchard is an artist-designer and geographer. As a practitioner they work together with others to make propositions and designs for computing otherwise, developing methods to uphold a politics of queer survival. Helen is an associate professor in queer feminist technoscience & digital design at i-DAT, University of Plymouth. They are the co-editor of *DataBrowser 06: Executing Practices* (2018) and *Science, Technology and Human Values: Sensors and Sensing Practices* (2019).

Blanca Pujals is an architect, spatial researcher and critical writer. Her cross-disciplinary practice uses spatial research and critical analysis to engage with questions around the geographies of power on bodies and territories, policies of scientific and technological knowledge production, as well as transnational politics, developing tools for undertaking analysis through different visual and sonic devices. Her work encompasses film, architecture, lecturing, curatorial projects, teaching and critical writing.

Jara Rocha is an interdependent researcher-artist. They are currently involved in several disobedient action research projects, such as Volumetric Regimes (with Femke Snelting), The Underground Division (with Helen Pritchard and Femke Snelting), The Relearning Series (with Martino Morandi), and Vibes & Leaks (with Kym Ward and Xavier Gorgol). They are part of the curatorial teams of DONE at Foto Colectania, of ISEA at Arts Santa Mònica and of La Capella, all in Barcelona; Jara also teaches screen studies at the Escola Superior de Cinema i Audiovisuals de Catalunya, as well as at the Körper, Theorie und Poetik des Performativen Department at Staatliche Akademie der Bildenden Künste, Stuttgart. With Karl Moubarak and Cristina Cochior, they conform the Cell for Digital Discomfort at the 21/22 Fellowship for Situated Research of BAK, Utrecht. Jara works through the situated, mundane, and complex forms of distribution of the technological with an antifascist and trans*feminist sensibility, and their show "Naturoculturas son disturbios" emits erratically from dublab.es radio.

Sina Seifee, born in Tehran 1982, is an artist based in Brussels. Using storytelling, video, and performance, he explores and teases with the heritage of zoology in West Asia. His work picks up on how epistemologies, jokes and knowledges about animals get shaped in the old and new intersections of techno-media and globalism. His work has been been presenting internationally in WIELS, Brussels (2020); SAVVY Contemporary, Berlin (2016); Sharjah Art Foundation, UAE (2018); Haus der Kulturen der Welt, Berlin (2017); Temporary Gallery, Köln (2019); Hordaland Kunstsenter, Bergen (2019); and Akademie der Künste der Welt, Köln (2015). http://www.sinaseifee.com

Femke Snelting develops projects at the intersection of design, feminisms, and free software in various constellations. With Seda Gürses, Miriyam Aouragh, and Helen Pritchard, she runs the Institute for Technology in the Public Interest. With the Underground Division (Helen Pritchard and Jara Rocha) she studies the computational imaginations of rock formations, and with Jara Rocha, Femke activates Possible Bodies. She is teammember of Programmable Infrastructures (TUDelft), i-DAT (University of Plymouth) and supports artistic research at PhdArts (Leiden), MERIAN (Maastricht) and a.pass (Brussels). Femke teaches at XPUB (MA Experimental Publishing, Rotterdam).

Spec is a structure based in Brussels structured around Pierre Huyghebaert, Sophie Boiron and several independent collaborators. The studio works mainly in the cultural, associative or public field. If its structure is of a commercial type, its work is not a commodity, but aims at the production of meaning. Speculoos is also part and works with Atelier cartographique. http://speculoos.com

The Underground Division is a collective research project on techniques, technologies and infrastructures of subsurface rendering and their imaginations/fantasies/promises. It is dug by Helen V. Pritchard, Jara Rocha and Femke Snelting with the help of many other others. Which are the presences, latencies, absences and potentials that need to be accounted for, in relation to that deep and thick complexity? The Underground Division bugs contemporary regimes of volumetrics that are applied to extractivist, computationalist and geologic damages. The research will eventually culminate in the trans∗Feminist Rendering Program, a hands-on situation for device making, tool problematizing and "holing in gaug". http://ddivision.xyz

Kym Ward lives and works in Bidston, Liverpool, UK. She is one of the founding hosts of the Bidston Observatory Artistic Research Centre, a not-for-profit study centre, focused on providing artists, writers, academics, performers, etc., with a cheap, temporary place to dictate their own working methods. She moves between more solitary performance research practice and organising and enabling alternative or non-hierarchical educations. Her interests lie in productive critique: of softwares' production of social relation, of technologies of organisation and, when possible, the cheeky reappropriation of institutional structure.

Item Index

Items from the *Possible Bodies Inventory* featured in *Volumetric Regimes*